WHAT WE'RE FIGHTING FOR *NOW* IS EACH OTHER

WHAT WE'RE FIGHTING FOR *NOW* IS EACH OTHER

Dispatches from the Front Lines of Climate Justice

Wen Stephenson

BEACON PRESS
BOSTON

Beacon Press
Boston, Massachusetts
www.beacon.org

Beacon Press books
are published under the auspices of
the Unitarian Universalist Association of Congregations.

18 17 16 15   8 7 6 5 4 3 2 1

This book is printed on acid-free paper that meets the uncoated paper ANSI/NISO specifications for
permanence as revised in 1992.

Text design by Ruth Maassen

Portions of this book have appeared in different form as essays and features in the *Nation*, the *Boston
Phoenix*, *Grist*, the *Boston Globe*, and *Slate*.

Excerpt from "Manifesto: The Mad Farmer Liberation Front." Copyright © 2012 by Wendell Berry
from *New Collected Poems*. Reprinted by permission of Counterpoint.

Excerpt from "The Heart's Counting Knows Only One" from *The Lives of the Heart* by Jane Hirshfield. Copyright © 1997 by Jane Hirshfield. Reprinted by permission of HarperCollins Publishers.

Excerpts from *The Book of Job*, translated and with an introduction by Stephen Mitchell. Copyright
© 1979 by Stephen Mitchell. Revised edition copyright © 1987 by Stephen Mitchell. Reprinted by
permission of HarperCollins Publishers.

Excerpts from *Riprap and Cold Mountain Poems*. Copyright © 2009 by Gary Snyder. Reprinted by
permission of Counterpoint.

Excerpts from "The Mountain Spirit." Copyright © 1996 by Gary Snyder from *Mountains and Rivers
Without End*. Reprinted by permission of Counterpoint.

Excerpt from "After Bamiyan." Copyright © 2014 by Gary Snyder from *Danger on Peaks*. Reprinted
by permission of Counterpoint.

Library of Congress Cataloging-in-Publication Data
Stephenson, Wen.
  What we're fighting for now is each other : climate justice and the struggle for a livable world / Wen
Stephenson.
     pages cm
  Includes bibliographical references.
  ISBN 978-0-8070-8840-1 (hardcover : acid-free paper) — ISBN 978-0-8070-8841-8 (ebook)
  1. Climatic changes—Political aspects. 2. Global warming—Political aspects. 3. Environmentalism. I.
Title.
  QC903.S74 2015
  363.738'74—dc23
                                    2015010976

For my children,
and all of our children,
and their children
after them.

Let your life be a counter-friction to stop the machine.

—HENRY DAVID THOREAU, "CIVIL DISOBEDIENCE"

CONTENTS

PREFACE

This is really happening.

The Arctic and the glaciers are melting. The oceans are rising and acidifying. The corals are bleaching, the great forests dying and burning. The storms and floods, the droughts and heat waves, are intensifying. The farms and savannahs are parched and drying. Nations are disappearing. People are dying. Mass extinction is unfolding. And all of it sooner and faster than science predicted. The window in which to prevent the worst scenarios is closing before our eyes.

And the fossil-fuel industry—which holds the fate of humanity in its carbon reserves—is doubling down, economically and politically, on all this destruction. We face an unprecedented situation—a radical situation. It demands a radical response. A serious response.

This is a book about waking up. It's about waking up, individually and collectively, to the climate catastrophe that is upon us—truly waking up to it, intellectually, morally, and spiritually, as the most fundamental and urgent threat humanity has ever faced. And it's about some of the remarkable, wide-awake people I have come to know and at times worked alongside—those I think of as new American radicals—in the struggle to build a stronger movement for climate justice in this country, still the most powerful, morally accountable, and indispensable nation on Earth. A movement that's less like environmentalism and more like the human-rights and social-justice struggles of the nineteenth and twentieth centuries. A movement for human solidarity.

Of course, any book like this must begin by acknowledging the science and the sheer lateness of the hour—the fact that, if we intend to address the climate catastrophe in a serious way, our chance for any smooth, gradual transition has passed. We must acknowledge the fact that without immediate action at all levels to radically reduce greenhouse emissions and decarbonize our economies—requiring a society-wide mobilization and a thus-far unseen degree of global cooperation, leading to the effective end of the fossil-fuel industry as we know it—the kind of livable and just future we all want is simply inconceivable.

The international community has committed to keeping the global temperature from rising more than two degrees Celsius (3.6 Fahrenheit) above the preindustrial average—the level, we're told, at which catastrophic warming can still be avoided (we've already raised it almost one degree Celsius, with still more "baked in," perhaps half a degree, within the coming decades). But there's good reason to believe that even a rise of two degrees will set in motion "disastrous consequences" beyond humanity's control—as former top NASA climatologist James Hansen, now at Columbia University's Earth Institute, and seventeen coauthors concluded in a December 2013 study. Catastrophic warming, by any humane definition, is virtually certain—indeed, already happening. Because even in the very near term, what's "catastrophic" depends on where you live, and how poor you are, and more often than not the color of your skin. If you're one of the billions of people who live in the poorest and most vulnerable places on the planet, from Bangladesh to the Sahel to Louisiana, even one degree can mean catastrophe.

But the world's climate scientists and leading energy experts, including the International Energy Agency in its *World Energy Outlook* reports, are telling us that unless the major economies drastically and immediately change course—decisively shifting investments away from fossil-fuel extraction and infrastructure, leaving all but a small fraction of proven reserves in the ground over the next three to four decades—we are headed for a global temperature rise of four or five or even *six degrees* Celsius (11.8 F) within *this century*. Kevin Anderson, former head of the UK's Tyndall Centre for Climate Change Research, has noted that even a rise of four degrees (7.2 F) would bring consequences "incompatible with an organized global community." The World Bank warned in its major 2012 report, prepared by Germany's Potsdam Institute for Climate Impact Research, that four degrees is likely beyond our civilization's ability to adapt—and "must be avoided."

But we're *not* avoiding it. That's the message of the world's climate scientists in the latest assessment report of the UN's Intergovernmental Panel on Climate Change (IPCC), issued in 2013 and 2014. We're plunging headlong toward the worst-case scenarios—critical global food and water shortages, rapid sea-level rise, social upheaval—and beyond.

In the summer of 2012, according to the National Oceanic and Atmospheric Administration (NOAA), approximately 80 percent of Arctic sea ice, measured by volume, was gone. As my colleague Mark Hertsgaard has reported, scientists estimate that we could have an ice-free Arctic summer as soon as 2020 or 2030. But keep in mind, the speed at which the Arctic has melted, thus far, greatly exceeds what scientists had the stomach to predict. And once the Arctic sea ice is gone, well, we don't really know what that will mean for the climate—but it's not good. Given the positive warming feedback of a heat-absorbing open ocean—and the potentially massive release of greenhouse gases from melting permafrost—it's safe to say all bets are off. And the ice-free Arctic, along with the accelerated melting of Greenland's great ice sheet, amounts to only one of several "tipping points" for the global climate system, which would render the effects of climate change irreversible on a human timescale. Scientists reported in 2014 that the great West Antarctic ice sheet appears to be collapsing, carrying an eventual eleven feet of sea-level rise, and it now appears that East Antarctica's enormous Totten Glacier is headed in the same direction, with another eleven or more feet in store.

Even without crossing such thresholds, the unimaginable is becoming all too imaginable. A 2013 study from researchers at the Potsdam Institute concluded that with warming of only three degrees Celsius—the low end of what's predicted for this century on our present course—12 percent of the global population could well face "absolute" water scarcity and 24 percent "chronic" scarcity. In other words, on our current trajectory, well within this century, more than a third of the human beings on this planet could face a catastrophic lack of water. Last I checked, you need water to grow food. Pointing to a September 2014 study on the increasing risk of "megadrought" in the United States by researchers at Cornell, the University of Arizona, and the US Geological Survey, climate expert Joseph Romm asked, "For the kind of Dust-Bowlification caused by a megadrought, what does the word 'adaptation' even mean?" (We're talking to you, California.) "Human adaptation to prolonged, extreme drought is difficult or impossible," Romm wrote in a 2012 piece for the journal *Nature* surveying the scientific literature on drought. "Historically, the primary adaptation to dust-bowlification has been abandonment; the very word 'desert' comes from the Latin desertum for 'an abandoned place.' . . . Feeding some 9 billion people by mid-century in the face of a rapidly worsening climate may well be the greatest challenge the human race has ever faced."

Of course there's uncertainty about exactly how these consequences will play out. There will always be uncertainty in anything as complex as climate science. But as MIT's Kerry Emanuel, one of the country's leading climate scientists (and a former Republican), likes to say in his public talks, "Uncertainty doesn't translate into 'no worries, mate.'" In fact, he says, it's the opposite: uncertainty "is a double-edged sword." It's possible, Emanuel and his colleagues acknowledge, that the impacts of climate change will be less severe, and arrive more slowly, than the most sophisticated models predict. But those are averages, which means it's *equally probable* that the impacts will be *more severe,* and arrive *much faster,* than predicted. So far, mounting evidence, like the rapid melting of the Arctic and Antarctic, suggests that the latter may be the case.

The question now is not whether we're going to "stop global warming" or "solve the climate crisis"; it is whether humanity will act quickly and decisively enough to salvage civilization itself—in any form worth salvaging. Whether any kind of stable, humane, and just future—any kind of just society—is still possible.

We know that if the governments of the world actually wanted to address this situation, in a serious way, they could. Indeed, a select few, such as Denmark and Germany, have begun to do so. (Denmark already produces nearly half of its electricity from renewables, and Germany close to 40 percent; both are moving aggressively toward the goal of 100 percent renewable energy by 2050.) Stanford's Mark Jacobson published an influential 2009 report outlining a path to 100 percent renewable energy globally, and in 2014 created state-by-state roadmaps for the United States to be fully powered by renewables by mid-century. The IPCC, the IEA, and many others are telling us that it is likely still possible, technically and economically, to hold warming to the two-degree limit—but *only* if the world takes the necessary action *now.*

The point, these experts want us to understand, is that the barriers are not technological or financial—they're political. And in the United States, without which there can be no effective global action, this is largely the result of a successful decades-long effort by the fossil-fuel industry and those who do its bidding (as shown by scholars like Harvard's Naomi Oreskes and Drexel University's Robert Brulle, as well as the public record of lobbying and campaign funding) to sow confusion, doubt, and opposition—obstructing any policies that might slow the warming, or their profits, and buy us time.

Let's be clear about what the preceding statement really means: given what we know and have known for decades about climate change, to deny the science, deceive the public, and willfully obstruct any serious response to the climate catastrophe is to allow entire countries and cultures to disappear. It is to rob people, starting with the poorest and most vulnerable on the planet, of their land, their homes, their livelihoods, even their lives—and their children's lives, and their children's children's lives. For profit. And for political power.

There's a word for this: these are *crimes.* They are crimes against the earth, and they are crimes against humanity.

Where does this leave us? What is the proper response? *Remain calm, we're told. No "scare tactics" or "hysterics," please. Cooler heads will prevail. Enjoy the Earth Day festivities.*

I'm sorry, the cooler heads have not prevailed. It's been a quarter-century since the alarm was sounded. The cooler heads have failed.

If you want sweet, cool-headed reason, try this: masses of people—most of them young, a generation with little or nothing to lose—physically, nonviolently disrupting the fossil-fuel industry and the institutions that support it and abet it. Getting in the way of business as usual. Forcing the issue. Finally acting as though we accept what the science is telling us—and as though we actually care about our fellow human beings.

Isn't that a bit extreme?

Really? Extreme? Business as usual is extreme. Just ask a climate scientist. The building is burning. The innocents—the poor, the oppressed, the children, your *own* children—are inside. And the American petro state—which, under the "all of the above" energy policies of Barack Obama, has overtaken Saudi Arabia as the largest producer of oil and gas on the planet—is spraying fuel, not water, on the flames. That's more than extreme. It's homicidal. It's psychopathic. It's fucking insane.

This is hard. Coming to terms with the climate catastrophe is hard. It's frightening. It's infuriating. It's heartbreaking. A friend of mine, a young woman you'll meet later in this book, says that it's like walking around with a knife in your chest.

And so I ask again, in the face of this situation—in the face of despair—how does one respond?

Rather than retreat into various forms of denial and fatalism and cynicism, many of us, and especially a young generation of activists, have reached the conclusion that something more than merely "environmentalism," and virtuous green consumerism, is called for. That the only thing offering any chance of averting an apocalyptic future—and of getting through what's already coming with our humanity intact—is the kind of radically transformative social and political movement that has altered the course of history in the past. A movement like those that have made possible what was previously unthinkable, from abolition to civil rights.

On September 21, 2014, some three hundred thousand people converged on the streets of Manhattan for the historic People's Climate March, demanding serious climate action from world leaders meeting at the United Nations two days later, a summit convened by UN Secretary General Ban Ki-moon to prod those leaders toward a global agreement at Paris in December 2015. I was there that day with my wife and our two children, along with many of my friends and colleagues in the grassroots climate movement—and it was thrilling, as we were joined by hundreds of thousands of people in New York and in cities around the world, the single largest day of climate demonstrations ever.

One of the slogans for the march was, "To change everything, we need everyone." And I couldn't agree more. That's what this book is about. But here's what would really change everything: first acknowledging that the mainstream, Washington-focused environmental movement—and the mainstream, Big Green "climate movement" that grew out of it—has failed. That we've already lost the "climate fight," if that means "solving the climate crisis" and saving some semblance of the world we know. That it was lost before it began—because we started so late. That it's time now to fight like there's nothing left to lose but our humanity.

And yet where does the courage and commitment and sacrifice required for that kind of fight—for the kind of radical movement we need—actually come from?

What I have found, in the stories of those profiled here, and many others, is that the climate struggle, like so many struggles of the past, is essentially a *spiritual* struggle—it forces us to confront the deepest, most difficult questions about ourselves. The climate catastrophe is so fundamental that it strikes to the root of who we are: it's a radical situation, and it requires a radical response. But not radical, necessarily, in the conventional sense of ideology. Rather, it confronts us with a kind of radical necessity—a moral necessity. It requires us to wake up—to face the facts, to find out who we really are—and to act. In some cases, to lay everything on the line: our relationships, our reputations, our careers, our bodies, maybe even our lives.

Historically, you could say that transformative movements arise from such an awakening—an awakening that you might call spiritual. But whatever you name it, this kind of awakening transforms individuals—and, sometimes, it transforms the world. To suggest that the kind of awakening I'm describing here might lead to such a transformation may seem fanciful. But it's our only hope.

~

While this book concerns itself with climate justice, it should be clear that I did not set out to write a history or survey of the climate-justice movement. The climate-justice movement in the United States is broad and diverse—including groups and networks from 350.org to the Climate Justice Alliance to the Indigenous Environmental Network, Rising Tide North America, and many others, including countless smaller, local and regional grassroots organizations. And the movement is of course global, with a history extending back to the early 1990s, at least, and the merging of local and regional environmental justice efforts with post-colonial and Indigenous human-rights and global-justice movements. I don't pretend to offer a comprehensive or representative account of it. The same goes for the mainstream climate movement, and the environmental organizations out of which it has grown. Those histories would require an entirely different book, or books, best written by someone other than myself.

Nor, I should add, is this book a "data-driven" macroanalysis, in which I presume to chart the one sure path forward, or prescribe any specific set of strategies and tactics. This is not a how-to manual for climate activists. (Though I do encourage you to try some of it at home.)

Instead, I've merely written from my own experience, my own personal journey into the climate movement in the United States, about a few of the people I've come to know who have committed their lives, and at times risked everything, to help inspire and build a more powerful and enduring movement in this country. People who have honestly confronted despair—a despair fully justified by our situation—and yet, somehow, have found the resolve to keep fighting. This book represents no more and no less than my own search for the moral and spiritual wellsprings of that kind of courage and commitment—and my search for what the very idea of "climate justice," at this late hour, might yet mean.

The prologue that follows is where that search begins.

Walking Home from Walden

Above all, we cannot afford not to live in the present. . . . Unless our philosophy hears the cock crow in every barn-yard within our horizon, it is belated. . . . There is something suggested by it that is a newer testament—the gospel according to this moment.

—HENRY DAVID THOREAU, "WALKING," 1851

The first time I walked to Walden, six miles north of my house in the thickly wooded suburbs west of Boston, it had nothing to do with the planet. And it didn't have much to do with Henry David Thoreau. I'd never been a communicant in the cult of Thoreau, never made my devotions in that temple. I'd done the assigned reading in college—"Civil Disobedience," obviously, and *Walden* (or parts of it)—and that was about all. It wasn't really my thing. And while I'd spent a lot of time in the Great Outdoors, imbibing sweet draughts of nature in what I thought was wilderness—in the mountains of California, where I grew up, and in Arizona and Utah, Idaho and Montana, New England and Nepal—by the age of forty, I'd somehow never read the line from Thoreau's essay "Walking," those words most quoted by environmentalists and nature writers everywhere: "in Wildness is the preservation of the world." In any case, I wouldn't have had a clue what it meant.

No, that first time, in the summer of '07, it wasn't about the planet. It was about me. All about me—and the changes going on inside my head.

A year or so before, in the spring of 2006, when I was an editor at the *Boston Globe*, I was in the midst of what's known as a "personal crisis"—in my case, I was battling burnout and exhaustion and anxiety. I was drinking too much—way too much. And in my effort to regain "balance"—actually to save my life, and to save my children, my son who was then six and my daughter who was two, from incalculable suffering—I not only decided to quit drinking, and to get help, but I started meditating. Tentatively at first, and desperately, I sat on a cushion in the early mornings, before the sun was up, and I reexplored the Zen and Taoist classics that I'd first dipped into in my twenties, backpacking around India and China. I won't pretend that I understood everything, or that I was doing it "right," or became any sort of adept. It was private, experimental, yet sincere and diligent. And I'm still at it. I practice at a Zen center in nearby Lexington.

It's impossible to exaggerate how far Zen Buddhism is from the Christianity, conservative and evangelical, in which I was raised—my incongruous Bible Belt upbringing in the seventies and eighties in suburban LA, where I was born, after my family moved to California from Texas. That's right, Texas. My parents and my whole extended family, on both sides, are from Texas—mostly the rural, working-class, small-town parts. My parents and sisters remain devout Christians, and we're still close. I may be estranged from their evangelical faith, but not from them. How can I put it? We differ on ways and means, but I empathize with their deepest yearnings—it's hard to shake the feeling that I'm in need of salvation.

It was also around this time, in 2006 and 2007, that I started walking on weekends—taking relatively short walks along trails in the nearby woods and open fields. The walks were a kind of therapy, a kind of drug— or better, another kind of meditation. I felt my senses coming back to me, as if from someplace far away, and I experienced—why should I be embarrassed to say this?—what I can only call an awakening, or reawakening, at once sensory and, yes, spiritual (an unjustly derided word). In any case, *something* was happening. Whatever it was, it was real.

Yes, I know this is all a cliché. But maybe you know what I mean— maybe you've stepped off into a sunlit meadow on a morning in late autumn, and stood there, dumbstruck, watching frost melt on a blood-red leaf, just you and the meadow. Or heard the crack of ice beneath your boot in winter echo like a rifle shot through the woods, startling birds into flight against the glow of the sky at dusk. Or stopped in your tracks in late summer as a great blue heron took wing only yards in front of you.

So these walks, you could say, were getting serious—and I looked to extend their range. Instead of driving a mile or two and parking at the entrance to a trail, why not really use my legs and *go somewhere*? Not just for exercise, but for the sake of walking, of seeing, of hearing. I wanted full immersion. I wanted to get saved.

I often drove past Walden Pond on my way from Wayland, the town where I live, to Concord—in fact it was one of the places I occasionally parked and walked the trails. Why not walk there from my house? Six miles is walking distance, I said to myself, just north from Wayland through the southwest part of Lincoln, to the Concord line and the pond. Why not walk to Walden?

Of course, it sounded hokey, ripe for parody—*Suburban man discovers nature and walks to Walden. In other news, Arctic ice recedes at a record pace* . . . But I didn't care. This wasn't about the planet and it wasn't about Walden. It was about the walk itself. It was about liberation—my private footloose adventure into the not-so-wild. If only for a morning.

And so I walked: through the leafy suburban streets at dawn looking for a fix, past the organic farm fields, the crops and livestock, protected woods and wetlands, historic conservation land—the rolling wooded landscape, broad bright clearings, hayfields, ponds and swamps, wind-blown wildflower meadows. A pastoral, suburban paradise.

I don't remember many details of that first walk, but I remember striding beside an open field in morning sun, air fresh and damp with dew, cool breeze before the heat of the day, elemental colors—yellows of wildflowers, greens of full foliage, blue sky and white clouds—and an exhilaration, an energy, pumping through my legs, my entire body, moving, breathing in the open air, alive and awake in the world.

Of course Walden itself, the modern reality of the place, was a buzz-kill. Morning joggers chatted loudly on the trails, swimmers churned the water in caps and Speedos, an early tour bus unloaded. I'd been to

Walden any number of times, joined the crowds of leaf peepers streaming out of the city when we lived in Cambridge. Just another tourist site, is how I thought of it, picturesque but hardly breathtaking. (Isn't it supposed to be bigger, and, you know, *wilder*?) In fact, as you circle the pond you're never quite out of earshot of a major commuter path—Routes 2 and 126 are not much farther than a stone's throw, and the MBTA tracks, the same railroad line Thoreau knew, skirt the northwest shore.

None of which really bothered me—I'd expected all that. It was the walk home that I hadn't counted on.

If the walk to Walden was liberation, the walk home was work. Footsore from all the pavement, the breeze no longer so fresh and cool as the sun climbed, by the time I stepped back onto my driveway, sweat soaked and thirsty, it was almost midday, and the weekend had resumed its relentless slide into the week. My transcendental trip dissolved back into the realities, and unrealities, of daily life.

But the memory, the sense of liberation, stayed. There was no denying I'd entered the surrounding landscape—or at least one six-mile stretch of it—in some entirely different way, and it had entered me, taken possession. If the trails close to home were what first reawakened me to "nature," now the vista opened wide, the perception deepened. I experienced it with almost a child's awe and wonder. It was a tactile reality, a place my own feet could carry me—and all I had to do was wake up, walk out, and touch it.

In the late 1950s in Berkeley, California, a young American poet named Gary Snyder translated some of the poems of Han-Shan, or Cold Mountain—the quasi-mythical Tang dynasty hermit, named for his mountain retreat, a hero of Chan (Zen) Buddhism, and perhaps its greatest poet. As Snyder wrote, "He is a mountain madman in an old Chinese line of ragged hermits. When he talks about Cold Mountain he means himself, his home, his state of mind." In my walks, lines from one of Han-Shan's best-known poems, as Snyder rendered them, lodged in my mind:

I settled at Cold Mountain long ago,
Already it seems like years and years.

Freely drifting, I prowl the woods and streams
And linger watching things themselves. . . .
Happy with a stone underhead
Let heaven and earth go about their changes.

Heaven and earth could go about their changes. I'd found my way to
Cold Mountain. It was out there, in a clearing, beside a pond, in a cloud's
reflection, on the way to Walden.

———

It was only later, three years later, in the spring and summer of 2010, that
the idea of walking to Walden, and home from Walden, took on new
meaning. Something fundamental had changed—in the country, in the
world, and in me. It dawned on me gradually at first, while out on my
walks, not to Walden but closer to home.

The place I'd walk, and still do, more than any other—a kind of sacred spot—is about a mile from my house, along my route to Walden,
where the street turns to gravel and dirt as it doglegs around a parcel of
protected woods running up a low ridge. The woods are bordered on
the north side by the small Hazel Brook and Stone's Pond. And sloping
gently down toward the pond from a dirt road and a low, rustic stone
wall is the most beautiful hayfield I'd ever seen—edged all around by big
graceful oaks, elegant maples, and tall, sturdy white pines. I've come to
know this view of field, pond, and woods as well as any place on Earth.
I've seen it and walked it in all seasons, in almost every kind of weather.

But as I skirted that field, down toward the pond, on a cold day in
May, there was something ominous in the air, which many could feel. The
drumbeat of reports on climate change, from scientists and journalists,
had grown more insistent, more alarming, over the past three years—
while the inability of our political establishment to meet the threat, despite the promises of Barack Obama, had grown alarmingly obvious.

We could now see with virtual certainty that global warming would
bring vast, and likely catastrophic, consequences—not least for human
beings, beginning with the poorest, most vulnerable, and least culpable—

within this century, our children's lifetimes, quite possibly our own life-times. The science told us that it was too late to stop it—given the ever increasing buildup of greenhouse gases, the amount already in the at-mosphere, and the sheer momentum of the climate system, we'd already guaranteed "dangerous" levels of warming in the coming decades. If we'd acted sooner, and decisively, we might have stood some chance of averting this—but we didn't. No matter what, we would now have to live through climate change, and to the extent possible, adapt to a very different Earth. The question now was whether we could still preserve a livable planet for future, and even current, generations.

But apparently the rising tide of evidence wasn't alarming enough. Anyone who followed the news could see the political disaster unfold in real time: as the economy trumped all other issues, we watched world leaders fail to act in Copenhagen; saw Congress failing to pass even weak bipartisan climate legislation; saw the environmental movement fail to mobilize the grassroots; saw public opinion in fact shift in the opposite direction, toward doubt and denial, as a party beholden to the carbon lobby, and openly dismissive of climate science, mounted a political re-surgence; saw the president of the United States—who as a candidate had spoken, in epic terms, of slowing the oceans' rise—fail to lead.

As these realities, scientific and political, sank in, it seemed that some kind of hinge point in history had been reached. I knew—indeed it was clear to anyone who cared to look—that despite the efforts of scientists and experts to warn us, for more than two decades, we had failed to act in time. I say *we*. I mean the elites, political and corporate and cultural, the privileged few, on the left as well as the right and center—and journalists, myself included, as much or more than anyone—because we knew, or should have known, what was happening. And yet we pretended to our-selves, in defiance of all reason, that somehow business as usual, politics as usual, life as usual, could go on. Pretended that an existential threat somehow called for something less than an existential response.

And so, there I am, one day in May 2010—at the edge of a field, walking toward a pond—and all these facts I've papered over, all this knowledge I've held at bay, can no longer be ignored. I can no longer pretend. It hits me, as though bodily, with a kind of visceral force, and

I know: *this place is already condemned.* In the blink of an eye, perhaps within my own lifetime, it will no longer exist. Not like this. And not because some future builder and bulldozer will destroy it, but because they—we—I—already have, by what we've already done, by what we've left undone. Striding through a hayfield on a cold, bright, and gusty New England morning, it can be hard to believe that the Arctic is melting, the oceans acidifying, the great forests dying, ancient glaciers disappearing. But I know that all of it is true, and that this sanctuary where I stand, this refuge, is a private delusion, a self-indulgent fantasy. There is no refuge. There is no sanctuary. Not for me, not for anyone.

And not only *this* place, but *all* places, for all time, all possible futures—condemned. The vanity, the emptiness, of all former assumptions, all former plans, all former striving, is exposed, naked. Time itself flies off a cliff, and the sense of anything solid, anything lasting, dissolves, melts into air. The future, my own children's future, vanishes, goes blank—and reappears as a nightmare, a time-lapse film flashing across the imagination. The landscape withers, burns, turns to ashes, dust—blows away. I'm in freefall. There is nothing to hold onto, nothing on which my future life, my children's lives, can stand. There's nothing but the ground beneath my feet, and this very moment, the moment in which I breathe.

I stop at the edge of the pond, see the reflection of trees and sky erased by the wind. And in the sudden void where my future self once stood—a choice, thrown back at me like an echo across the water: What will I do with the time I'm given?

My sheer belatedness in reaching this point is mind bending—in fact it's perhaps the most difficult thing for me to understand and accept. How could I have gone so long in denial? Not denial of the science, the fact of climate change—I was always reasonably well informed. I mean denial, on some deeper level, of my own part in it, my responsibility—personal and political—and of what climate change will mean in the years to come.

I took no comfort in knowing that I was far from alone in my lateness. My entire generation, more or less, as we entered middle age, stood indicted. Those of us born in the late sixties and early seventies came of age along with the knowledge of global warming. And yet, a full twenty years on, here we were, the vast majority of us, having done little or nothing about it.

I, for one, couldn't be bothered—I had a digital revolution to win. I had degrees to earn, a career to build, a family to start. I had places to go, websites to launch. I had bands to see, blogs to read. I had endless cable TV dramas to watch. I had drugs to take, coffee to grind, weight to lose. I had *memoirs* to write. For Christ's sake, *I had my navel to observe*. You could say I navel gazed while the planet burned.

What I'm trying to say here, what I believe needs to be said, is that what I experienced that day at Stone's Pond—the fear, the anguish, the palpable sense of an existential reckoning, all breaking to the surface— was a symptom of something far larger than my own private struggle. What I'm saying is, there's a *spiritual crisis*, or struggle, at the heart of the climate crisis and the climate struggle—a crisis we've hardly begun to come to grips with, or even acknowledge. The immense suffering that is now inevitable, within this century, on this rapidly warming planet is the result not only of an "environmental" or "economic" or "political" crisis—or even, for that matter, a "moral" one. It's all of these combined, and yet, if possible, more. It's what I can only call spiritual.

By "spiritual" I don't necessarily mean religious (although often it is). I mean deeply *human*—I mean our deepest, most profound, and ultimately inexpressible sense of ourselves. And by "crisis" I mean a deep crisis of identity unlike any, perhaps, since Darwin, or Hiroshima—an unnerving sense that, despite all our science and technology, we don't really know who we are and where we're going, or what it means to be alive as a human being at this moment on Earth. A sense that we don't yet know the full magnitude of what we've done to the planet and to future generations, beginning with our own children. A paralyzing sense that we're heading into the unknown, into a new, uncharted wilderness for which we're not prepared. And I want to say that it's our society's failure, our failure, on the whole, to acknowledge and address this essentially spiritual dimension of our global crisis, humanity's greatest, that explains our failure to come to grips with it, morally as well as politically.

What I'm talking about transcends "environmentalism." It transcends religion. It transcends politics and ideology. What I'm talking about is the overriding fact of our shared human predicament. And what I want to suggest is that it requires something of us beyond the usual politics and proposals, the usual answers, the usual pieties. It requires a kind of searching, a kind of questioning, that can move us from self-absorption, isolation, cynicism—from the moral and emotional detachment of denial—to a new kind of engagement.

Or maybe a very old kind.

I don't know what it was, exactly—some voice on the wind?—but standing there at Stone's Pond, something drew me back to Walden and Thoreau. I'd gotten to know my surrounding physical landscape, but I was lost when it came to the moral one, and Thoreau is that interior landscape's most prominent local feature—the height looming over my shoulder. Hadn't he written about the moral crises of his time, slavery and war and industrialization, in the grip of a passionate personal engagement with nature? How could I have managed to ignore him? He'd lived just up the road, our local saint, walked the same ground. It was as though I'd been living at the foot of Whitney or Washington—or Katahdin—and never thought to climb it.

And so I decided, as ridiculous on some level as it might sound, that the first thing I'd do was go back to Thoreau—and I would walk back to Walden. Only this time, I'd pay more attention—I'd find new routes. I'd look for the other Walden, Thoreau's Walden. I had to see if it could still be found. Not the place itself, but the state of mind. And not a refuge or sanctuary, but a way of living in the world, of engaging it, as it is, right now. That's what I had to find—that and the answer to the only question now that mattered, the question Thoreau had asked himself and all of us: *What will we do with the time we're given?*

———

Coming back to Thoreau after twenty years, one of the first things I noticed (and I'm hardly the first) is just how deeply religious, or spiritual, his relationship to nature—that is, to everything, the world—actually is. I knew that he'd dipped into Eastern religions, mainly India's, and was steeped in Greek paganism, but I never thought of Thoreau as a religious writer. Some would rather keep him squarely in the secularist camp, hostile to Christianity in particular. And it's true that Thoreau was no Christian in any orthodox sense. But it seems to me his dispute was with the church and the pulpit of his day—and the severely constricted view of God and salvation they allowed—not the notion of the divine itself, or for that matter, the ethical teachings of Jesus. On his deathbed, asked by a devoutly Calvinist aunt whether he'd made his peace with God, Thoreau replied, "I did not know we had ever quarreled."

Of course, when you think about Thoreau, you may be saying, "Yeah, sure, Walden Pond, nature, climate change. I get it. Wasn't he basically our first environmentalist?" Well, actually, no. And that's the other thing I realized coming back to Thoreau. Sure, he was a great naturalist, maybe even our first ecologist—but Henry David Thoreau, icon of the American environmental canon, was not an "environmentalist" (a term that would have made no sense to him). He was, however—without question—a *radical abolitionist.* He was what we would call a human-rights activist. In fact Thoreau's great subject, in *Walden* and just about everything he wrote, wasn't "the environment" or even "nature." It was how to live as a human being in relation to both nature and *other human beings*, because the two can't really be separated.

What would it mean, then, for us, at this moment, if Thoreau can't be understood without grasping both the deep spiritual impulses that drove him and the radical abolitionist impulses? And what if that great essay "Walking"—arguably Thoreau's central essay, because it's so closely related to *Walden*—isn't just an essay on the secular meaning and value of "wildness" but is also, maybe more importantly, a kind of sermon? A sermon on discovering not only the source of individual liberation, but of *human liberation*—and the source of something like *solidarity*—in "the wild"?

In "Walking"—first delivered as the lectures "Walking" and "The Wild" in 1851, when *Walden* was still three and a half years from publication, the draft set aside, languishing in a proverbial drawer—Thoreau makes the case for waking up to our immediate surroundings, our physical and moral landscape, both human and wild. He not only wants "to speak a word for Nature, for absolute freedom and wildness," as he states in that famous first line, but just as much, "to regard man as an inhabitant, or a part and parcel of Nature."

The stakes are high, as Thoreau makes clear at the outset—nothing less than our souls. "They who never go to the Holy Land in their walks . . . are indeed mere idlers and vagabonds." Even more, "If you are ready to leave father and mother, and brother and sister, and wife and child and friends," he writes, alluding to the Gospels and the story of Jesus calling the Apostles, "then you are ready for a walk." And why leave all behind and venture in search of the Holy Land—if not for salvation?

The walk, then, becomes a sort of pilgrimage. But that still leaves the question, what kind of salvation is Thoreau preaching?

"In my walks," Thoreau writes, "I would fain return to my senses." That is, wake up to what's going on, at this very moment, all around him, whether a farm field, a woodlot, a swamp—or his own body and mind, his society and its laws, his own conscience. He had already written "Civil Disobedience," and was active in the Underground Railroad. It's all interconnected: humanity, the farm, civilization, all depend on the wild—the source of it all. And what his senses reveal—when he returns to them—*is* the wild, the wild-*ness*, in everything, himself included. "Life consists with wildness. The most alive is the wildest."

This spiritual aspect of wildness, for Thoreau, is inherently social and political. "I enter a swamp as a sacred place—a *sanctum sanctorum*," Thoreau writes. "There is the strength, the marrow of Nature." But he goes there not to escape. Indeed, for all his emphasis on nature, Thoreau never takes his eye off of society, the town, and its reform. "A town is saved," he writes—yes, *saved*—"not more by the righteous men in it than by the woods and swamps that surround it. . . . and out of such a wilderness comes the Reformer eating locusts and wild honey"—an allusion to John the Baptist and the Hebrew prophets before him. A town, in the end, is saved—reformed—by wildness.

Which brings me back to this radical preacher's most radical statement: "in Wildness is the preservation of the world." I don't know, but something tells me this is more than a bumper sticker for conservation. If *wildness* is our true source, if that's where we find our true selves, then to realize this wildness *within* is to find what Thoreau calls God—and to find salvation, liberation. Not in some faraway Heaven, some other-worldly eternal life, but here, now. And not only liberation for *oneself*—in some self-absorbed desire to get saved—but for everyone, everything, the *world*. Because you and the world are not separate. In that realization is "the preservation of the world." And what is *preservation*? Obviously, to protect or save, as in conservation. But also to save from going rotten or stale, to keep fresh, ripe. The preservation of the world is the preservation of ripeness—wildness, readiness, rightness—within ourselves, in the present moment.

The present moment is key. It's where you are when you return to your senses. It's what your senses return you *to*. "Above all, we cannot afford not to live in the present," Thoreau writes. "Unless our philosophy hears the cock crow in every barn-yard within our horizon, it is belated. . . . There is something suggested by it that is a newer testament—the gospel according to this moment."

"Walking" was among the last pieces of writing Thoreau prepared for publication, on his deathbed, not long before he died in April 1862. "So we saunter toward the Holy Land," he writes at the end, "till one day the sun shall shine more brightly than ever he has done, shall perchance shine into our minds and hearts, and light up our whole lives with a great awakening light. . . ."

Wake up, walk out, and be saved.

If you understand "Walking," you can almost skip *Walden*. (Actually, please don't.) What I mean is, it's clear that "Walking," and the actual walking that inspired it—Thoreau's long excursions into the surrounding countryside—led to *Walden*. Within a year of delivering the lecture for the first time, in the spring of 1851, Thoreau was back at his draft of the big book, revising and expanding with renewed creative energy. You could almost say Thoreau "walked to *Walden*."

And yet if "Walking" is a sermon, then *Walden* is something more like prophecy—its author the Reformer and child of wildness, divine messenger, sent to save the town. He'll issue his call, he boasts at the outset, "as lustily as chanticleer in the morning, standing on his roost, if only to wake my neighbors up."

More even than "Walking," *Walden* is a wake-up call. The sonorous second chapter, "Where I Lived and What I Lived For," centers on a kind of hymn to morning, to daybreak, "the awakening hour"—the hour of re-creation and rebirth. The morning, he tells us, "is when I am awake and there is a dawn in me. . . . To be awake is to be alive." And as with wildness, something about morning and sunrise is strongly linked, for Thoreau, with the idea of the present—and the sacred. "God himself," he writes, "culminates in the present moment, and will never be more divine in the lapse of all the ages."

But this awakening, this renewal, none of it means a thing if it doesn't lead to action—to reform, individual and social. To live a single day "as deliberately as Nature," Thoreau is saying, would be to wake up and be liberated to act on one's conscience. "Moral reform," he writes, "is the effort to throw off sleep."

There's a popular image of Thoreau as a recluse, aloof and detached, even a little misanthropic—a crank indulging his private fantasy in his hideaway in the woods. But Thoreau's cabin at Walden was never intended as a retreat or refuge from the world or society. As Thoreau himself describes in *Walden*, he not only kept up a social life at the pond, he remained socially and politically engaged. He spent a night in Concord's jail for refusing to support an imperialist proslavery government. The Antislavery Society met in his cabin. Indeed, if anyone took refuge there in Walden's woods (if not in Thoreau's cabin), it was the runaway slaves Thoreau helped along the Underground Railroad to freedom.

More than some kind of anachronistic environmentalism, it's precisely Thoreau's antislavery activism, his radical advocacy of what today we call human rights, that should be remembered as central to his legacy—a legacy that is every bit as much Gandhi and Martin Luther King as the Nature Conservancy.

Thoreau understood slavery as the moral and spiritual crisis of his time. Opposing it with words and actions was a clear moral imperative—and a clear directive of nature. As he wrote in "Walking," where we hear the philosophy of the cock crow, "no fugitive slave laws are passed." In fact, the Fugitive Slave Law was very much on Thoreau's mind as he delivered the "Walking" lecture for the first time: an 1851 draft shows him opening with a reference to the escaped slave Thomas Sims, who had been captured in Boston and sent back to Georgia that very April.

In May 1854 another runaway slave named Anthony Burns was cap-tured in Boston, setting off a conflagration of protest. An abolitionist mob, led by the radical Vigilance Committees—organized by the likes of Theodore Parker and Thomas Wentworth Higginson together with African American ministers and other members of the city's black community—made a dramatic attempt to free Burns from the courthouse by force. Only with the intervention of state and federal troops on the streets of Boston was Burns sent back into slavery in early June.

Weeks later, on July 4, 1854, with *Walden* in final page proofs, Thoreau mounted a platform at Harmony Grove in Framingham—alongside William Lloyd Garrison, Sojourner Truth, and other prominent abolitionists—to address a fiery antislavery rally. (Literally fiery: Garrison put copies of the Fugitive Slave Law and the US Constitution to the torch.) The speech Thoreau delivered, "Slavery in Massachusetts," is merciless in its contempt, as he lacerated the commonwealth for the moral complacency and hypocrisy of its participation in human bondage. It was enough to shake his sense of nature's harmony:

> I walk toward one of our ponds, but what signifies the beauty of nature when men are base? . . . Who can be serene in a country where both the rulers and the ruled are without principle? The remembrance of my country spoils my walk.

And yet, in the closing moments of the speech, recalling one of his recent walks through the landscape around Concord, he finds reassurance:

> But it chanced the other day that I scented a white water-lily, and a season I had waited for had arrived. . . . What confirmation of our hopes is in the fragrance of this flower! I shall not so soon despair of the world for it, notwithstanding slavery, and the

cowardice and want of principle of Northern men. It suggests what kind of laws have prevailed the longest and widest, and still prevail, and that the time may come when man's deeds will smell as sweet. Such is the odor which the plant emits. If Nature can compound this fragrance still annually, I shall believe her still young and full of vigor, her integrity and genius unimpaired, and that there is virtue even in man, too, who is fitted to perceive and love it. It reminds me that Nature has been partner to no Missouri Compromise. I scent no compromise in the fragrance of the water-lily . . .

For Henry Thoreau, to live in harmony with nature is to act in solidarity with one's fellow human beings.

———

That summer of 2010, a so-called natural disaster half a planet away made excruciatingly clear what it means to be alive as a human being at this moment on Earth. On July 22, as rains fell on Pakistan like nobody had ever seen, the rivers began to flood, eventually inundating one-fifth of the country, among the poorest in Asia, submerging seventeen million acres of its most fertile land, devastating its infrastructure, and leaving at least five million people shelterless, thousands dead, and some ten million still in "urgent need," as the *New York Times* reported, by early September. All in a politically unstable, nuclear-armed country at war on two fronts.

On the far side of the world, another story was being written, on what might seem an entirely different planet—at high elevation in the Sierras of California and the mountains of Nevada and Utah, the rim of the North American Great Basin—and on a completely different scale. A story on the front page of the *Times* that September told us that the bristlecone pines—the world's oldest living trees, some dated at nearly five thousand years—appear now to be threatened by climate change. A combination of "blister rust" fungus from Asia and the devastating pine bark beetle (the same one ravaging the forests of the Rockies), aided by warmer winters and drought, now makes the bristlecone's fate uncertain. Living trees older than the world's great religions, as old as the civilization born in Pakistan's inundated Indus River valley, a species adapted to the harshest imaginable climate and soil—seemingly far beyond the touch of human culture—now inexorably succumbing to our carbon exhaust.

These two images, from the timberline of the high Sierras to the floodplains of Pakistan, said it all—what I'd sensed that day at Stone's Pond, only orders of magnitude larger. The stories and footage from Pakistan—at the very least, portents of global warming's coming effects, if not a result of climate instability already observed—and the fate of those five-thousand-year-old pines, pointed to the true nature of our crisis: not merely "environmental," but *human*. This planet- and civilization-altering force is none other than ourselves. But not only that, the truth is double edged: global warming, human caused, is now increasingly human suffered.

At his own very different moment—when the coal-fueled, industrial economic forces now warming the planet were young—Thoreau had a run-in with incomprehensibly "vast, Titanic, inhuman nature." On the summit of Maine's Mount Katahdin in September 1846, Thoreau famously confronted, for the first time, a landscape "savage and awful. . . . no man's garden, but the unhandselled globe." And yet, crucially, he also confronted his own place in that landscape, his identity with those forces, right down to the elements composing his own body, a shock of recognition he couldn't fully assimilate:

> I stand in awe of my body, this matter to which I am bound has become so strange to me. . . . Talk of mysteries!—Think of our life in nature,—daily to be shown matter, to come in contact with it,—rocks, trees, wind on our cheeks! the *solid* earth! The *actual* world! The *common sense! Contact! Contact! Who* are we? *Where* are we?

We're there with Thoreau on Katahdin, even as his questions—*Who are we? Where* are we?—take on new meaning and urgency. Thoreau struggled to conceive of his own material identity with wind, rocks, and trees. And yet he wasn't forced to confront, as we are now, his own complicity in the destruction of Earth's ecological balance, even the atmosphere. He didn't have to make sense of his own part in a force felt from the crags of the high Sierras to the rivers of South Asia, source waters of the civilization that gave him his dear *Bhagavad Gita.* But that's the landscape in which we now stand.

I finished *Walden*—all of it this time—in early August, out on our screened porch late at night. Turning the last pages, I heard the sound of insects in a darkness made all the deeper by my dim lamp. If you didn't know there were other houses all around, you could almost convince yourself you were in a cabin in the woods.

At the end of *Walden,* Thoreau returns to his theme of awakening and rebirth. In the penultimate chapter, "Spring," he finds the soil coming back to life and discovers "there is nothing inorganic. . . . The earth is not a mere fragment of dead history . . . not a fossil earth, but a living earth." He's become an ecologist, but the impulse, and the deeper meaning of the season, is still a spiritual and moral one. "In a pleasant spring morning all man's sins are forgiven. Such a day is truce to vice. . . . Through our own recovered innocence we discern the innocence of our neighbors."

In the final chapter, the ringing Conclusion, it's those forgiven neighbors he still addresses, even if his message is too radical for them to grasp. "I desire to speak somewhere *without* bounds; like a man in a waking moment, to men in their waking moments . . ." And yet most people, Thoreau knew, would not see what he saw, if only because the truth can be blinding. "The light which puts out our eyes," he writes, "is darkness to us." (In the New Testament, I recall, Saint Paul is briefly blinded by a divine light on the road to Damascus.) But the light of morning is no less real for all of that, Thoreau seems to say, if we can only rouse ourselves to see it. The final words of *Walden* are these: "Only that day dawns to which we are awake. There is more day to dawn. The sun is but a morning star."

~

Five days later, the weather hot and clear, I got up with the sun and walked again to Walden. I'd researched and scouted the conservation land in Lincoln, had my maps of the trails, and I wanted to walk there without following the major roads—to keep on the trails as much as possible. The route was circuitous, far more than six miles this time, but I was in full-on Thoreau pilgrimage mode, heading for the Holy Land.

At Codman Farm I left the road and entered the woods, then cut across the big hayfields along Route 126 (what Thoreau called the "Wayland road"), and crossed the street near Saint Anne's in the Fields, a small Episcopal church. There I entered the Mount Misery conservation area, cut through the small Lindentree Farm at its northeast corner, southwest along a large open meadow, then north into the woods and onto a public easement trail through secluded tracts of private property, and in sight of a few backyards, until I crossed over Heywood Brook and entered the large parcel called Adams Woods, bordering Walden Woods on the south and west.

It was quiet, I was all alone on the trail, and off to my left, a few hundred yards to the southwest, was Fairhaven Bay, where the Sudbury River widens out into a lake of sorts, a favorite of Thoreau's and the local waterfowl (you can still find the latter). By now it was midmorning, and the sun was slanting through the pines and lighting up the leaves of birch and oak, like some impressionist stained-glass window. I stopped and looked up into the branches—felt their hush, a faint breeze in the treetops. Cathedral hour.

I came out onto Walden, crossing the railroad tracks at the far opposite end of the pond from the busy visitor center and beach. A few people were fishing quietly along the bank as I headed around the shore toward the site of Thoreau's cabin, above the inlet at the north end of the pond. There were the swimmers, as always, making their long laps like ritual bathers in the clear, sacred water. It occurred to me that Thoreau might well be happy to know that the shores of his beloved pond offer the closest thing I've ever seen in America to the ghats on India's Ganges. True, they're not exactly the "burning ghats" of Varanasi—no cremations here, strictly bathing. But I can't help the thought that Thoreau should have been cremated on these banks, the boards of his cabin for a pyre, his ashes spread on the water with garlands of marigolds and floating candle flames . . .

We'd had a long, dry summer, and walking home I was conscious of the climate. I'd read of invasive species, seen some of them, aided by the weather; of the dry summers stressing our waterways, of more insects surviving the shorter winters and now threatening our forests. Of songbirds disappearing. Of wildflowers Thoreau knew never returning.

On the way back I kept to the main roads, mostly. It was the same landscape I'd reveled in three years before—and had trudged over on that first tiring walk home. But now it looked different—a lot less like paradise and a lot more like the suburbs.

Now, along with the earth-friendly farms, I noticed the houses. Endless houses, most of them big, some of them enormous—and not a few of those brand-spanking new—with enormous garages for their enormous vehicles. I thought of those quiet streets lined with cars in a few hours, the evening commute. I thought of the fact that even here, in environmentally progressive Massachusetts, a majority of voters favored the two challengers for governor who spoke of pulling out of the modest Regional Greenhouse Gas Initiative, and put tax cuts ahead of investments in clean energy (the incumbent, Deval Patrick, went on to win reelection, but with less than 50 percent of the vote). I thought of the fact that the commonwealth had sent a climate denier and servant of the fossil-fuel lobby to the United States Senate.

But who am I to preach to you? And why should you listen? How am I all that different, really, than those sleepwalking into the climate future? I'm certainly no Thoreau. I'm not even an environmentalist. One of those houses is in fact mine—not the largest in the neighborhood, far from it, but large enough, with two cars in the garage (not SUVs, but not hybrids either), two kids, a flat-screen TV, the usual gadgets. You get the picture. Enough to contribute my share of America's outsized greenhouse emissions. Sure, my house now has a rooftop full of solar panels—but so what? Are they there to salve my conscience? Do they buy me absolution? Who the hell am I?

Only a privileged citizen of the most powerful nation on Earth. Only one of those on whom everything now depends.

"The remembrance of my country spoils my walk."

I'm sorry, Henry, but I don't believe that. I don't believe it merely spoiled your walk. The remembrance of your country revealed the walk's true purpose. It's not about some solitary quest for the Holy Land, some private back-to-nature trip; it's about the journey out of self-absorption to engagement. That's what you're telling me. Because you won't let us forget—will you, Henry?—that your awakening in nature led you back to society and a radical political engagement in solidarity with your fellow human beings.

I step back onto my driveway, pick up the *New York Times* and the *Boston Globe* in their plastic wrappers, walk around to the porch and collapse in the chair where I'd finished reading *Walden* just a few nights before. I know that walking to Walden—some idea of nature or God or some private salvation—isn't enough. It's what you bring home, what you leave behind. It's what you *do* when you get home.

I walk out into my own backyard. The sun slants through pines. Gospel hour.

"Only that day dawns to which we are awake. There is more day to dawn."

This is Walden. Wake up. Start here.

The sun is climbing the sky.

PART ONE

ONE

The New Abolitionists

The scientists found that to hold warming to 2C, total emissions cannot exceed 1,000 gigatons of carbon. Yet by 2011, more than half of that total 'allowance'—531 gigatons—had already been emitted.

—FIONA HARVEY, "IPCC: 30 YEARS TO CLIMATE CALAMITY IF WE CARRY ON BLOWING THE CARBON BUDGET," *The Guardian*, SEPTEMBER 27, 2013

If I have unjustly wrested a plank from a drowning man, I must restore it to him though I drown myself . . .

—HENRY DAVID THOREAU, "CIVIL DISOBEDIENCE," 1849

I want to say a word for the radical—for the role of a reasoned and morally serious radicalism in the struggle for climate justice at this critical moment, this do-or-die moment, in human history. I want to begin by quoting some famous words, indeed some of the most radical words ever spoken by an American.

On August 3, 1857, speaking in Canandaigua, New York, on the twenty-third anniversary of the emancipation of the British West Indies, the abolitionist Frederick Douglass offered "a word," as he put it, "of the philosophy of reform." Douglass's philosophy was clear, direct, fierce—even, as Henry Thoreau might have said, wild. It was the kind of "newer testament" that Thoreau wanted to preach himself—the kind that "hears the cock crow."

"The whole history of the progress of human liberty," Douglass told the assembled crowd, "shows that all concessions yet made to her august claims, have been born of earnest struggle." He went on, in what has become perhaps his most quoted, if little heeded, passage:

> If there is no struggle there is no progress. Those who profess to favor freedom and yet deprecate agitation, are men who want crops without plowing up the ground, they want rain without thunder and lightning. They want the ocean without the awful roar of its many waters. This struggle may be a moral one, or it may be a physical one, and it may be both moral and physical, but it must be a struggle. Power concedes nothing without a demand. It never did and it never will.

Frederick Douglass, a man who was born a slave and who escaped—who *freed himself* from slavery—knew something about struggle. But my ancestors on this continent were never enslaved. I've never suffered racial or any other kind of oppression. I've never had to fight for any fundamental rights. All I want is a livable world—and the *possibility* of social justice. So who am I to invoke Frederick Douglass?

I'll tell you who I am: I'm a human being. I'm the father of two young children. I'm a husband, a son, a brother, a citizen. And I am engaged in a struggle—a struggle—for the fate of humanity and of life on Earth. Not a polite debate around the dinner table, or in a classroom, or an editorial meeting—or an Earth Day picnic. I'm talking about a *struggle*. A struggle for justice on a global scale. For human dignity and human rights for my fellow human beings, beginning with the poorest and most vulnerable, both far and near. For my own children's future—but not only my children, all of our children, everywhere. A life-and-death struggle for the survival of all that I love. Because that is what the fight for climate justice is.

And so I want to say a word for the radical—and for the kind of radical movement that has made possible, politically and socially, things that were previously unthinkable. The kind of movement that responds to a situation that is itself radical—radically threatening, radically unjust—in which people are forced to decide who they really are, and what their conscience requires of them, and what they must do.

I want to begin with two scenes, and two very different speakers, who embody the imperatives, and the limitations, of the moment in which we find ourselves.

July 26, 2011. Inside a federal courtroom in Salt Lake City, Utah, a twenty-nine-year-old climate activist named Tim DeChristopher is sentenced to two years in prison and a $10,000 fine for disrupting a Bureau of Land Management auction of oil and gas drilling leases back in December 2008. Registered as Bidder 70, he had managed to win bids worth $1.8 million for some twenty-two thousand acres of public land near Canyonlands National Park—bids he had no way of paying. DeChristopher had acted spontaneously, on his conscience, engaged in nonviolent resistance to the heedless, and undemocratic, new extraction of fossil fuels that are catastrophically heating the planet and threatening innumerable innocent lives. (The auction itself, a kind of sweetheart parting gift to the industry by the outgoing George W. Bush administration, was invalidated in early 2009 because the environmental review process had been found insufficient, and leases covering more than one hundred thousand acres were cancelled. But DeChristopher was prosecuted nonetheless.)

Weeks before his sentencing, DeChristopher—who cofounded the group Peaceful Uprising in Salt Lake City shortly after his action—tells *Rolling Stone*'s Jeff Goodell: "I'm a climate-justice activist. . . . We want a radically different world. We want a healthy, just world." But first, he says, "we need to get the fossil fuel industry out of the way. First we've got to overthrow the corporate power that is running our government." He understands what that requires. "It will involve confrontation and it will involve sacrifice."

That day at his sentencing, standing before the federal judge, De-Christopher delivers a long, eloquent statement that spreads across the Internet and galvanizes a growing climate-justice movement. "This is not going away," he says in conclusion. "At this point of unimaginable threats on the horizon, this is what hope looks like. In these times of a morally bankrupt government that has sold out its principles, this is what patriotism looks like. With countless lives on the line, this is what love looks like, and it will only grow."

A month after DeChristopher speaks those words, the largest wave of civil disobedience in a generation—organized by 350.org and Tar Sands Action—begins in front of the White House, where 1,253 activists are arrested protesting the Keystone XL pipeline, the project that would carry eight hundred thousand barrels per day of Alberta tar-sands crude from Canada to the Gulf Coast for export, accelerating the extraction of one of the largest carbon deposits on Earth.

November 4, 2012. Inside the town hall in Arlington, Massachusetts, on the Sunday before Election Day, one week after Hurricane Sandy made landfall, Congressman (now US Senator) Edward Markey stands before a capacity crowd. Hundreds of constituents have gathered, on forty-eight-hours' notice, for what the congressman billed as an "emergency meeting" on climate change. Markey displays satellite photos of Boston illustrating that huge sections of the city—like the entire Back Bay—would be underwater if Sandy had hit the Hub instead of New York and New Jersey.

But Markey isn't there just to talk about disaster response or building sea walls in Boston Harbor. He's there to demonstrate his seriousness on confronting climate change, an issue that had until that week gone all but unmentioned in the presidential election campaign and in the mainstream political media (indeed, even in elite, left-leaning media).

"As the Minutemen responded, so must we," Markey tells his audience, calling for an unspecified "bold plan" from Washington to cut greenhouse emissions and prevent future "devastation." Global warming, if unaddressed, could lead to "events so horrific," he says, that they could "dwarf" other catastrophes in human history.

In his final remarks Markey intones, with what sounds like real passion: "The American Revolution, it started here. The abolitionist movement, it started here. The women's movement, it started here. The anti-Vietnam movement, it started here. . . . The Freedom Riders, going south in the sixties, they left on buses from here. . . . [Global warming] is our generational challenge. The preceding generations accepted their challenges."

~

I was there at Arlington Town Hall that Sunday in November 2012, and I had to wonder: If Ed Markey was as serious as he sounded about climate change, what kind of "bold" action would match the challenge of the moment and his rhetoric invoking the grand radical tradition in American history?

Certainly nothing that he or any other politician in Washington has ever proposed, including (especially) President Barack Obama, comes anywhere close. Even the doomed 2009 cap-and-trade bill that Markey coauthored with Rep. Henry Waxman of California—the only comprehensive national climate legislation ever to pass either chamber of Congress—aimed merely to cut emissions 17 percent below 2005 levels by 2020. That's the same amount, as it happens, that Obama meekly pledged at the failed UN climate talks in Copenhagen that same year. Compare that with what the international scientific consensus at the time, as represented by the IPCC's 2007 *Fourth Assessment Report*, said would be required if humanity is to have a chance of stabilizing the climate: at least 25 to 40 percent below *1990 levels* by 2020. In other words, using the internationally recognized 1990 baseline, the Obama-Waxman-Markey target would amount to a roughly *4-percent* emissions reduction by 2020.

And yet almost without fail, mainstream media, including those to the left of center, have referred to Obama's policies to meet that pledge—whether auto-emissions standards or EPA regulations on power plants—as "aggressive." Even Obama's widely hailed handshake agreement with China, announced in November 2014, in which he increased his pledge to 26–28 percent below 2005 levels *by 2025*, and in which China pledged to stop *increasing* its emissions *by 2030*, was called "ambitious" and a "landmark." Which it was, considering that not even such a toothless deal as that one had ever existed before. Meanwhile, under Obama, the United States has become the world's largest producer of oil and gas.

Around the time that Markey spoke in Arlington, the International Energy Agency (IEA), the World Bank, and PricewaterhouseCoopers (PwC)—not exactly environmental organizations—were releasing reports that would surely have been described as "alarmist" if issued by climate advocates. (As it happened, the reports went virtually unmentioned by major news outlets, including the *New York Times*.) The generally conservative IEA affirmed that at least two-thirds of proven fossil-fuel reserves must stay in the ground, between then and 2050, in order to have a shot at keeping the global average temperature from rising more than two degrees Celsius, the internationally agreed-upon "red line." Meanwhile, as we've seen, the World Bank warned that the planet is on track for a four-degree-Celsius rise this century—which it said "must be avoided." The analysts at PwC, in a report titled *Too Late for Two Degrees?*, concluded that we've "passed a critical threshold," and should prepare for four degrees, or even six degrees, this century, unless the carbon-intensity of the global economy can be reduced by an unprecedented 5 percent per year for the next forty years. (A more recent analysis by the Deep Decarbonization Pathways Project, a multinational research effort backed in part by the IEA, maintains that staying below 2C is still technologically and economically feasible—but assumes, among other things, immediate action by all countries and a truly unprecedented degree of global coordination. As climate analyst David Roberts of *Grist* magazine noted, "These are substantial assumptions!")

In July 2012, months ahead of the IEA's report, Middlebury College environmental scholar Bill McKibben, cofounder of 350.org, introduced the idea of a "carbon budget" to a mass audience in a landmark article for *Rolling Stone* headlined "Global Warming's Terrifying New Math." (The carbon budget concept, long understood by climate experts, was finally endorsed by the IPCC in its 2013–14 *Fifth Assessment Report*.) Drawing on a new analysis of fossil-fuel reserves by the London-based Carbon Tracker Initiative, a group of climate and financial experts, McKibben spelled out the bottom line in cold, hard arithmetic: *565 gigatons* = the amount of CO_2 scientists agree we can still pump into the atmosphere and hope to remain below the two-degree threshold. *2,795 gigatons* = the amount of CO_2 contained in the world's proven fossil-fuel reserves, which the fossil-fuel industry shows every intention of extracting and burning. In other words, the then-known reserves were five times larger than a sane carbon budget would allow. We have to find a way to leave *80 percent* of accessible fossil fuels in the ground, *forever*, while making an all-out effort to shift to clean energy, in the next three to four decades. (The IEA's estimate that two-thirds of reserves must stay in the ground is based on a cold-blooded 50 percent probability of staying at or below two degrees; McKibben and the Carbon Tracker analysts would prefer slightly better, and more humane, odds.)

It seems fairly obvious that the reason we don't hear politicians, or the "serious" people in our media, talking much about this situation—the true gravity of it—is that to grapple with this in any real way, to propose anything that would actually begin to address it with the necessary urgency at the national and global level, would simply sound too extreme, if not outright crazy, within mainstream political conversation. *Leave fossil fuels in the ground? You must be joking. That would mean leaving the Alberta tar sands untapped! It would mean shutting down coal plants and canceling coal-export terminals! It would mean no new fossil-fuel infrastructure—not even natural gas! It would mean no fracking! Who are you kidding? Be serious.*

This is the reality—or the *surreality*—of the historical moment in which we find ourselves. At this late hour, with the clock ticking down on civilization, to be *serious* about climate change—based, mind you, on what science and not ideology prescribes—is to be *radical.*

———

In drawing historical comparisons between the climate movement and radical struggles for justice and human rights, Ed Markey was echoing the sentiments of a good many climate activists and movement leaders. And it seems that, for many of us, the comparison that most deeply resonates is to abolitionism: the stunningly radical and successful movement, led by a small yet fervent minority first in Britain and then the United States, to abolish the legal institution of human slavery on which a large part of the global economy was based.

Much of what appeals about the analogy is the bracing moral clarity and uncompromising urgency of the abolitionist cause. In 1831, introducing the first issue of the *Liberator* in Boston, William Lloyd Garrison answered his moderate critics: "I do not wish to think, or to speak, or write, with moderation. No! no! Tell a man whose house is on fire to give a moderate alarm . . . tell the mother to gradually extricate her babe from the fire into which it has fallen;—but urge me not to use moderation in a cause like the present. I am in earnest—I will not equivocate—I will not excuse—I will not retreat a single inch—AND I WILL BE HEARD."

"We need the urgency of a William Lloyd Garrison, or even more," Bill McKibben told me in a 2012 interview, agreeing that the climate-justice movement, with its emphasis on human rights, has more in common with nineteenth-century abolitionism than with much of today's environmentalism.

There are significant caveats to the comparison, of course, just as there are to any historical analogy. Here are three big ones. First, there is no one-to-one equivalence between the consumption of fossil fuels to power our daily lives and put food on our tables (whether we're rich or poor) and the enslavement, systematic torture, and mass murder of countless human beings on the basis of race. Nothing compares with that. Second, it should go without saying that fossil fuels and their effects on the atmosphere cannot simply be abolished at the stroke of a pen. There will be no Emancipation Proclamation or Act of Parliament freeing us from fossil fuels, no constitutional amendment abolishing climate change.

Finally, the climate movement advocates and engages in strictly *nonviolent* resistance and protest. When it comes to direct action, its models are Gandhi and Martin Luther King, not John Brown. (The Garrisonian abolitionists, I should note, were staunchly nonviolent.) That's not to say that violence was unjustified in the struggle to end slavery—or that it would *never* be justified to save millions of lives from the effects of climate chaos. But the climate movement has always been a resolutely nonviolent movement.

What resonates, then, is not the analogy to slavery itself, or any literal comparison to abolitionist tactics, but the *role* of the abolitionist movement, *as a movement,* in American and human history—and the necessity now of a movement that is every ounce its morally and politically transformative equivalent.

The parallels are irresistible: There's the sheer magnitude of what's at stake, in human and moral and, yes, in economic terms—countless millions of lives and many trillions of dollars ($20 trillion in "unburnable" fossil-fuel reserves, according to one analysis cited by McKibben). There's the fierce opposition of powerful and entrenched reactionary forces—the Slave Power of the antebellum South and the fossil-fuel lobby of today. There's the movement's explicit emphasis on human rights and social justice, including economic and racial justice, considering that the vast majority of those suffering the earliest and worst impacts of climate change globally are impoverished people of color—who, let's be clear, have done nothing to create the catastrophe and whose economic development and very survival now depends on a massive deployment of renewable energy that will be impossible without significant assistance, the payment of a "climate debt," by those wealthy industrialized nations that have, in fact, created the catastrophe.

And then there are the spiritual underpinnings of both movements, the progressive religious inspiration of many activists and leaders: abolitionism grew out of Quakerism, early evangelicalism, and African American churches, while today's climate-justice movement has deep support not only among progressive faith communities (as did, let's not forget, the civil rights and antiwar movements of the sixties)—and deeply spiritual Indigenous communities—but it's also the case that many secular, nonreligious activists are motivated and sustained by what they acknowledge is a kind of spiritual commitment.

The historian David Brion Davis, in *The Problem of Slavery in the Age of Emancipation* (2014), the final volume of his epic Problem of Slavery trilogy, writes that abolition "probably stands, despite the US Civil War and other heavy costs, as the greatest landmark of willed moral progress in human history"—the result of a "revolutionary shift in moral perception." If the abolition of slavery was the great human, moral struggle of the eighteenth and nineteenth centuries, then climate justice is the great human, moral struggle of our own time. And the climate-justice movement has every reason to be as resolute and as radical, in its own way, as the movement that ended slavery.

Now, I believe that the previous statement is true. In fact I've committed the rest of my life to it. And yet I also know that any proposition so large is never so simple. I know that history, and the nature of radicalism, are complicated. Climate justice may well be the greatest human-rights struggle of our time, but actions, however pure the motive, have consequences, and we need to be honest with ourselves about the consequences of radicalism, then and now—even as we're honest with ourselves about the consequences of not being radical enough.

In the spring of 2012, as it happens, a fresh debate cropped up over the meaning and legacy of the American abolitionist movement, thanks to a brilliant and provocative essay called "The Abolitionist Imagination" by Andrew Delbanco of Columbia University.

Delbanco is interested in American abolitionism not simply as a specific movement at a specific time and place in history, but as he puts it, "an instance of a recurrent American phenomenon: a determined minority sets out in the face of long odds to rid the world of what it regards as a patent and entrenched evil." The abolitionists, Delbanco notes, "belonged self-consciously to the tradition of imprecatory prophets; they were the thundering Isaiahs and Jeremiahs of their time, calling to account this fallen world and exploiting the fear of apocalypse if they should fail."

Viewed in this light, Delbanco goes on to ask whether abolitionism *should* be the model or inspiration for present-day justice and liberation struggles (although the climate movement goes unmentioned). He reminds us that, far from being admired as the morally fearless heroes we remember them as today, they were derided and reviled by their contemporaries. The word "abolitionism" was most often used as "a slander meant to convey what many Americans considered its essential qualities: unreason, impatience, implacability." Stephen Douglas compared his archrival Lincoln in 1858 to "the little abolitionist orators in the church and school basements." In 1860, Lincoln—no abolitionist, but an antislavery moderate who gradually came to accept abolition—distanced himself from the radical movement.

To be sure, Delbanco leaves no doubt about what the abolitionists achieved. "The contribution of the abolitionists was to make thinkable what had once been unthinkable, namely, black freedom," Delbanco writes, pointing to historian Eric Foner's assessment in his Pulitzer Prize–winning 2010 book *The Fiery Trial: Abraham Lincoln and American Slavery.* "By pushing beyond conventional ideas about race and slavery," Delbanco writes, "they changed both Lincoln's private judgment and public opinion, thereby vastly enlarging what was politically possible in nineteenth-century America." You won't find a better description of what the climate movement might hope to achieve—if in place of "race and slavery" you substitute fossil fuels and climate.

But Delbanco's major point, what some readers seem to find most provocative, is that it's entirely possible to give the abolitionists their full due, yet still sympathize with the "intellectual and political leaders who, although disgusted by slavery, nevertheless tried to forestall the catastrophic war they feared was coming."

Herman Melville described slavery as "a sin . . . no less—a blot, foul as the crater-pool of hell," but despaired that "not one man . . . knows a prudent remedy." Both he and his friend Nathaniel Hawthorne were re-pelled by the abolitionists' extremism because, it seems, they didn't want the blood of a cataclysmic war on their hands. "They sensed," Delbanco tells us, "that Armageddon was coming—and that, if abolitionists and fire-eating slaveholders had their way, it would come soon." Melville's mono-maniacal Ahab in *Moby-Dick* was seen as "a timely personification of the zealotry that was rising, in 1850–51, on both sides of the slavery divide."

Delbanco wants us to be alert and sensitive to this kind of moral complexity, and empathetic toward those who were sincerely conflicted about pushing too hard, too fast. The "sacred rage of abolitionism," he writes, "has been at work in many holy wars since the war against slav-ery." And so Delbanco would hold us back "from passing easy judgment on those who withheld themselves from the crusade, not out of indiffer-ence, but because of conscientious doubt."

I know where Delbanco is coming from. I've felt that same shud-der. Indeed, those stirring words of Frederick Douglass, quoted at the beginning of this chapter, contain an unambiguous threat of violence, both inflicted and suffered. It's worth noting what follows immediately after that passage in Douglass's speech. "Find out just what any people will quietly submit to and you have found out the exact measure of injustice and wrong which will be imposed upon them, and these will continue till they are resisted with either words or blows, or with both," Douglass says. "The limits of tyrants are prescribed by the endurance of those whom they oppress. . . . If we ever get free from the oppressions and wrongs heaped upon us, we must pay for their removal. We must do this by labor, by suffering, by sacrifice, and if needs be, by our lives and the lives of others."

Frederick Douglass was a former slave—many would argue that he, and all in his situation, had every right (even the duty, Douglass believed) to speak in terms of "any means necessary." And I'd like to think that if I'd been a contemporary of Hawthorne and Melville, I would have had the courage and clarity to be among the abolitionists. But the truth is, no matter how virtuous I want to believe myself, I simply don't know. Nor, if we're honest, do any of us.

I do know, however, what it is to care passionately and urgently about an issue, a *cause*, of enormous magnitude—political and moral, even spiritual—only to find myself at once attracted and repelled, fascinated and frightened, by the voice of a radical.

——

The first time I recall reading about Tim DeChristopher, it was in the spring of 2011, around the time of his trial. In the months between his conviction that March and his sentencing in late July, a number of stories and interviews popped up, and I came across a Q&A in the socialist UK magazine *Red Pepper*. "We are at a time in our movement," DeChristopher said there, "where we need to be honest"—that it's "too late to stop a climate crisis," and that averting unthinkable catastrophe will now require deep, urgent, transformative changes. "We are not looking for small shifts: we want a radical overhaul of our economy and society."

Now, you have to understand, I'm not exactly a lifelong lefty. I've never been much of a leftist at all. I spent two decades as an editor and producer in the mainstream media, where I considered myself a thoughtful, centrist independent. (I've never registered for any party.) I was heavily influenced, I admit, by Bill Clinton's winning triangulations (if not his deceptions). I'm a climate activist now—even a climate-justice activist—but with my house in the suburbs, my two young children, and my spouse with her MBA, I'm an unlikely radical, to say the least.

So when I read DeChristopher in *Red Pepper*, my first reaction was, "No. What are you doing? You can't say that stuff. This sort of talk, if it goes too far, has consequences. People are listening to you now. If the movement radicalizes, we'll alienate people, we'll be marginalized, we'll never get anything from Congress—we'll sacrifice genuine, if incremental, progress for the sake of some kind of moral, or ideological, purity. And we don't have time for that. We have to take whatever progress we can get."

I had only recently walked away from my media career, and was just beginning to get involved in the grassroots climate movement. I was still trying to fit my ideas of what needed to be done inside the suffocatingly cramped quarters of the politically "possible" at that moment. I had yet to fully face the facts of the situation in front of us.

But that fall, the news from the climate front was unrelentingly grim: global emissions set new records, extreme weather and melting ice caps showed accelerating climate impacts, the IEA told us we're on track to blow past the two-degree limit on our way to six degrees, Oxfam reported that climate change is already threatening global food security . . . and it went on. Meanwhile, a presidential campaign got going under the influence of the fossil-fuel funded Tea Party, pushing Republicans ever further into denial and obstruction, aided by a climate-silent media cowered into false "balance." It became clear that even modest, incremental steps—much less comprehensive, economy-wide national measures— were a pipe dream in Washington.

By late December, I bottomed out—in despair for the planet and my children's future. *We're fucked,* I realized. *Now what?*

More or less at that moment, Tim DeChristopher came back into view, in a long, astonishing interview with Terry Tempest Williams in *Orion* magazine, recorded the previous May as DeChristopher awaited sentencing. I see it now as an essential, primary document of the American climate-justice movement. And what happened, quite simply, is this: DeChristopher, a convict, convicted me.

In that interview, DeChristopher tells of the "shattering" moment in March 2008 when he met climate scientist Terry Root, a lead IPCC author, at a symposium at the University of Utah:

> She presented all the IPCC data, and I went up to her afterwards and said, "That graph that you showed, with the possible emission scenarios in the twenty-first century? It looked like the best case was that carbon peaked around 2030 and started coming back down." She said, "Yeah, that's right." And I said, "But didn't the report that you guys just put out say that if we didn't peak by 2015 and then start coming back down that we were pretty much all screwed, and we wouldn't even recognize the planet?" And she said, "Yeah, that's right." And I said: "So, what am I missing? It seems like you guys are saying there's no way we can make it." And she said, "You're not missing anything. There are things we could have done in the eighties, there are some things we could have done in the nineties—but it's probably too late to avoid any of the worst-case scenarios that we're talking about." And she literally put her hand on my shoulder and said, "I'm sorry my generation failed yours."

"Once I realized that there was no hope in any sort of normal future," DeChristopher goes on, "I realized that I have absolutely nothing to lose by fighting back."

DeChristopher expresses here what I had been repressing. He knows that building the sort of movement that can "fight back"—and create the conditions in which we can build, with a lot of luck and perseverance, a stable and ultimately better world—will require something of us beyond the ordinary conduct of politics. The climate crisis, he says, justifies "the strongest possible tactics in response," by which DeChristopher means "nonviolent resistance." That doesn't mean everyone has to go to prison, as he did. But, he says, "the *willingness for that* is what's necessary. That willingness to not hold back, to not be safe."

The willingness to not be safe.

"You can't 'move the center' from the center," DeChristopher goes on to say near the end of that interview, referring to Naomi Klein's often-quoted formulation that the movement's job is to move the political center. DeChristopher adds: "If you want to shift the balance—if you want to tilt that scale—you have to go to the edge and push. You have to go beyond what people consider to be reasonable, and push."

DeChristopher was released from federal custody in April 2013, and that fall he entered Harvard Divinity School, with plans to become a Unitarian Universalist minister (he had been a member of a strong and supportive Unitarian congregation in Salt Lake City). Since I live just half an hour west of the Harvard campus, I've now had a chance to get to know him, even collaborate with him as an activist, including on fossil-fuel divestment at Harvard (and you'll read about our conversations later in this book). But you can imagine how I felt when I learned that Tim was coming to Cambridge. *This guy won't let me hide.*

Tim DeChristopher is an abolitionist. And when I think about the ways in which his story and his words have affected me, I can only empathize with Andrew Delbanco's brief for "conscientious doubt." I know that DeChristopher can be a little scary. He scared the shit out of me.

But here's the rub: today, in our present situation, one can argue that those who will have the blood on their hands, those who will be judged most culpable by our children and future generations, are not only the denialists and obstructionists on the right, but the moderates, the cautious pragmatists—the reasonable, serious, centrist voices—who fail to acknowledge the true scale, urgency, and gravity of the climate catastrophe, and so fail to address it in any meaningful way.

People like that—and I was one of them—will say that people like DeChristopher have no "plan," no "workable solutions." But it's not Tim DeChristopher's or the climate movement's job to offer detailed policy prescriptions that fit within the confines of our current politics. Given our political deadlock, the movement's job is to tell the truth, however extreme—and to force those in power to recognize that even the *outer limit* of what our current politics will allow (a modest carbon tax, for example) is utterly *inadequate* to the crisis. The movement's job is to force that reckoning.

Yes, radicalism still carries risks, as it always has. But at a moment when political possibility is closed off—when any policy commensurate with the challenge is dead on arrival—we have to ask, does an insistence on radical honesty, like DeChristopher's, really risk anything meaningful at all? Politically, at least, what is there to lose?

Of course, you might say I'm understating the risks of radicalization, that there may be other real consequences, from the personal to the social: that friendships, marriages, families may be torn apart; jobs lost, careers ruined, life options foreclosed; that there will be economic hardship, including for those who can least afford it, only increasing the dire necessity of economic justice; that social unrest, even violence, may erupt (just ask anyone who remembers the sixties).

Yes, I understand.

What, then, is the alternative? Because the risks of moderation and incrementalism, of accepting and working within our current political constraints, are infinitely more grave. The risks of moderation are a matter of life, death, and suffering for untold millions of human beings, alive today and yet to be born. If we can't radically alter our politics—radically expand the limits of what's politically thinkable, as the abolitionists did in Douglass's day—then we might as well not even talk about "climate action."

We might as well change the channel, and drift back to sleep.

———

As the battle over Keystone XL intensified in 2012 and 2013—inspired by First Nations peoples' fights against devastating tar-sands extraction in Alberta, strip-mining operations so vast as to create scars across the boreal forest visible from space—it became the central rallying point for a rapidly broadening and diversifying climate movement at a pivotal and radicalizing moment. More and more, the "fossil fuel resistance," as Bill McKibben dubbed it, was turning to nonviolent direct action and civil disobedience to make its demands seen and heard.

The resistance spread across the country, with struggles intensifying against mountaintop-removal mining in Appalachia, coal export terminals in the Pacific Northwest, and shale-gas fracking in the Northeast. Along the KXL route through Montana, South Dakota, and Nebraska, ranchers and property owners joined with Native American groups defending sacred and sovereign land. Perhaps most dramatically, along the Keystone's southern leg from Cushing, Oklahoma, to the Texas Gulf Coast (fast tracked by Obama in a March 2012 speech in Cushing, a transparent election-year pander to the oil lobby), members of the Tar Sands Blockade and Great Plains Tar Sands Resistance—including climate-justice activists, property owners, Indigenous allies, and members of frontline communities—put their bodies in the way of the pipeline's construction, often at great physical and legal risk. (Out of their actions arose the term "Blockadia," which came to refer to direct-action campaigns resisting fossil-fuel extraction and infrastructure across North America.) In early March 2013, the progressive group CREDO Action issued a call to activists to resist the pipeline, and by May some sixty thousand people across the country had pledged to engage in peaceful civil disobedience if the Obama administration moved toward approval (as of early 2015, the number had reached ninety-seven thousand). The Sierra Club, for the first time in its 120-year history, officially lifted its ban on participating in civil disobedience. Its executive director, Michael Brune, was among 48 protesters arrested at the White House on February 13, three days before the attention-grabbing "Forward On Climate" rally, spearheaded by the Club and 350.org, when some fifty thousand people gathered on the National Mall to protest Keystone and demand serious climate action by the Obama administration.

When Brune announced the Sierra Club's decision that January, in a short, eloquent piece titled "From Walden to the White House," he explicitly invoked the legacy of Henry David Thoreau and, of course, Thoreau's most famous essay, "Civil Disobedience." For Brune, as for many other activists, myself included, engaging in civil disobedience is a sacred American tradition. (On October 7, 2013, I sat down along with thirty-five fellow activists at the entrance to the Thomas P. O'Neill Jr. Federal Building in Boston to protest Keystone XL. There was a heavy police presence, and the authorities warned us ahead of time that we risked federal charges and fines, but they chose not to arrest us.)

Even so, as the climate movement has embraced that legacy, it's worth taking a step back and remembering just how radical Thoreau really was—and to remember what it was, exactly, that made him so. Not his night in the Concord jail—that was the easy part—but something else: a readiness to speak the truth, forcefully and without compromise, no matter how fanciful or extreme it may have sounded to cynical ears or what risks it might have entailed. What's more, if we're going to invoke Thoreau, we should remember that the logic of "Civil Disobedience" led him directly, a decade later, to "A Plea for Captain John Brown."

If that thought doesn't make you pause, it should. Those, like me, who invoke Thoreau as we engage in building this movement should ask ourselves if we're really ready to walk in his footsteps—and what it might mean, at this radical moment, if we did.

I have to admit that I've never much liked "Civil Disobedience," the essay Thoreau began drafting in his cabin at Walden Pond in the fall of 1846. The tone is just a little too arch, his performance somewhat preening. "I was not born to be forced," he writes. "I will breathe after my own fashion. Let us see who is the strongest." Regardless of such posturing (or perhaps because of it?), you can't help feeling that there's not a whole lot at stake for him personally—that he was, in a way, slumming it there in jail for a night—so that the essay takes on the air of a privileged intellectual exercise, a kind of abstract thought experiment to be conducted, after a good dinner, in Mr. Emerson's parlor.

Still, for all the mannered poses, there's a reason the essay has lasted, that its influence extends across continents and centuries. So it's worth reminding ourselves what Thoreau is really saying in "Civil Disobedience." From a relatively minor incident, now wrapped in legend, Thoreau gets down to first principles. In the last week of July 1846, he was stopped on his way into town to get a shoe repaired and asked to pay his poll tax. He refused, even though it meant jail. The country was engulfed in controversy over the Mexican War, a flagrant act of aggression to expand slave territory to the west, and there was even secession talk in the North. But why, Thoreau wants to know, should he wait for a vote in the statehouse? "Must the citizen ever for a moment, or in the least degree, resign his conscience to the legislator? Why has every man a conscience, then?"

Pressed to decide how he will act toward "this American government," Thoreau determines, "when a sixth of the population of a nation which has undertaken to be the refuge of liberty are slaves"—and one's own country has unjustly invaded another—"I think that it is not too soon for honest men to rebel and revolutionize."

The moral equation, Thoreau is saying, isn't terribly complicated. There are expedient reasons to recognize the authority of a government, he admits. But he insists that we recognize those situations "in which a people, as well as an individual, must do justice, cost what it may." He goes on, in the very next lines, to offer a stark analogy: "If I have unjustly wrested a plank from a drowning man, I must restore it to him though I drown myself. . . . This people must cease to hold slaves, and to make war on Mexico, though it cost them their existence as a people."

From this straight-up, no-nonsense formulation, Thoreau lays down a marker, a point from which he'll navigate. "Action from principle," he tells us, "the perception and the performance of right, changes things and relations; it is essentially revolutionary, and does not consist wholly with anything which was. It not only divides States and churches, it divides families; ay, it divides the *individual,* separating the diabolical in him from the divine."

This is strong stuff—and prophetic, in more ways than one. What we have here is a kind of working definition of Thoreau's radicalism: call it the willingness to face the "essential facts" (as he put it in *Walden*), and then to act as both facts and conscience require. Doing so, he assures us, "is essentially revolutionary"—the only real way to change the world.

"Let your life be a counter-friction," Thoreau writes, "to stop the machine."

Thoreau delivered the lecture that became "Civil Disobedience" in 1848, but he didn't publish the essay (originally titled "Resistance to Civil Government") until the next year, when it was denounced as "crazy" and "radically against our government," practically French in its revolutionary zeal. It had been two years since he left his cabin by the pond, and his life had settled into a routine of long walks, writing (including his increasingly ambitious Journal), working as a surveyor, and occasional lecturing. But in the 1850s, events in Concord and beyond provided ample opportunity for "action from principle."

As we've seen, Thoreau's image as an apolitical hermit has always been a caricature; his active involvement in the Underground Railroad and resistance to the Fugitive Slave Act put the lie to it. We know that he helped multiple fugitives on their way to Canada, guarding over them in his family's house—the Thoreau family were committed abolitionists, especially his mother and sisters—even escorting them onto the trains, not without personal risk. In the fall of 1859, his principles would be put even further to the test.

Henry Thoreau met John Brown in March 1857, when the hero of "Free Kansas," already famous, or infamous, for his bloody exploits—today we would call them war crimes—came through Concord on a speaking and fundraising tour of the Northeast. Brown was invited to lunch at the Thoreau home on Main Street, where Emerson joined them in the afternoon. Thoreau and Emerson spent hours talking with Brown, sizing him up, and came away greatly impressed. Brown spoke at Concord Town Hall that night, and the Emersons hosted a reception.

But not everyone in Concord was so taken with Brown—far from it—and when the news arrived in October 1859 of Brown's deadly raid on Harpers Ferry, Virginia, reactions were sharply divided. The whole country was in an uproar. Even Brown's erstwhile supporters quickly distanced themselves. Most of his co-conspirators—many with close ties to Concord—went into hiding, several fleeing to Canada. The atmosphere was tense, even dangerous, for those voicing solidarity with Brown.

Into this picture steps forty-two-year-old Henry Thoreau, now a leading intellectual. Incensed by the timid and hypocritical reactions of his neighbors, and of the press, Thoreau let it be known that he would speak in support of Brown at Concord's First Church on October 30. Thoreau rang the town bell himself (Concord's selectmen had refused). The address he gave was "A Plea for Captain John Brown."

It was Thoreau's most radical moment. He was the first in Concord, and among the first and most prominent in the country, to come to Brown's defense. Within days he would repeat the speech to large audiences in Worcester and Boston—where he stood in at the last moment for Frederick Douglass, who had been chased into Canada by federal marshals despite having played no part in the Harpers Ferry raid.

The speech itself is stunning. What Thoreau was saying in his "Plea" for Brown was the same thing he'd said a decade earlier in "Civil Disobedience"—"action from principle . . . is essentially revolutionary"—only in far stronger terms, and this time with real skin in the game. What was once a kind of philosophical exercise was now in deadly earnest: Brown's raid and certain execution—not to mention the risk of publicly aligning oneself with him—made Thoreau's night in jail look like child's play.

But what I find most striking about Thoreau's "Plea" isn't simply the fact that he championed the violent and fanatical Brown (*religiously* fanatical, it's important to remember; Brown saw himself on a divine mission, engaged in a holy war against slavery). Rather, it's the rhetorical strategy Thoreau chose. Thoreau explicitly sets out to defend Brown not in the court of conventional opinion, nor of any state or constitution, but in the court of a "higher law"—of conscience. "I plead not for his life," Thoreau tells his audience, "but for his character—his immortal life." Most of all, and most profoundly, it becomes clear, this means pleading for Brown's *sanity.*

Nothing offends Thoreau more than the knee-jerk reaction among his neighbors, and even many abolitionists, to write Brown off as a madman. He has no patience for them: "They pronounce this man insane, for they know that *they* could never act as he does." But he saves his fiercest ridicule for fellow abolitionists, including William Lloyd Garrison's

Liberator: "Even the *Liberator* called it 'a misguided, wild, and apparently insane . . . effort,'" he writes. "Republican editors . . . express no admiration, nor true sorrow even, but call these men 'deluded fanatics'— 'mistaken men'—'insane,' or 'crazed.'" This pushes Thoreau over the edge: "Insane! . . . while the sane tyrant holds with a firmer gripe [sic] than ever his four millions of slaves, and a thousand sane editors, his abettors, are saving their country and their bacon! . . . Ask the tyrant who is his most dangerous foe, the sane man or the insane." Far from insane, Thoreau argues, Brown was the "superior man," even Christ-like—an explicit, if rather strained, comparison throughout the speech.

In defending not only Brown's actions but his sanity against the moderate opinion of what we might call the "center" and "center-left," Thoreau was pushing hard on the boundaries of acceptable discourse. He was, as the saying goes, moving the center. He forced his listeners to consider what was truly "sane" and "insane" in the face of slavery. For Thoreau, Brown's was a "saner sanity," recognizing the fact that slavery, intolerable on every level, would never be abolished in the United States without bloodshed. This is what it meant, Thoreau was saying, to be sane in America in 1859.

On December 2, Brown was hanged in Virginia. The next day, Thoreau himself would become an accomplice to the escape of a desperate Harpers Ferry conspirator, Francis Jackson Merriam, personally taking him out of Concord by wagon to the train in Acton. Thoreau didn't know Merriam's identity (he was told only to call him "Lockwood"), but he surely knew what he was doing and the risk he was taking—that this was a wanted man, with a price on his head.

"In my walks, I would fain return to my senses," Thoreau wrote, with characteristic wordplay, in "Walking." It's the same essay in which he wrote the line most quoted by conservationists: "in Wildness is the preservation of the world." In John Brown, Thoreau would encounter a human force of nature, a kind of wildness, that he hoped would bring the country to its senses, its sanity, on the question of slavery—the kind of sanity Thoreau had expressed in "Civil Disobedience": "This people must cease to hold slaves . . . though it cost them their existence as a people."

———

Fortunately for us, Thoreau—with his explicit endorsement of Brown's violence—didn't get the last word on civil disobedience (he was apparently free of the "conscientious doubt" that nagged at his friend Hawthorne). Instead, Mahatma Gandhi, the Reverend Dr. Martin Luther King Jr., and many others transformed resistance to intolerable injustice in ways Thoreau never imagined—demonstrating the power of a steadfast, principled, radical *nonviolence*. Don't let anyone tell you that "radical" is synonymous with "violent." Gandhi and King were the best kind of radicals. So was Jesus—whose nonviolence Thoreau conveniently omitted from his plea for Brown.

And yet today we face a human crisis as extreme in its way as the one faced by Thoreau. What, then, is the "sane"—and appropriately radical—response to the urgent human crisis of global warming? Is anyone willing to say, "This people must cease to extract and burn fossil fuels, and to unjustly rob today's children and future generations of a livable planet, whatever the cost"?

It sounds crazy. But just as Thoreau and other radical abolitionists were willing to push the boundaries, so the climate-justice movement, if we're serious, must not cower from saying and doing "crazy" and "radical" things—like disrupting federal auctions, or putting our bodies in the way of pipelines and coal shipments, or facing jail to demand that our universities divest from fossil fuels—not because it's politically expedient, but because it's morally imperative. When the truly sane courses of action—putting a heavy price on carbon, massively scaling up clean energy, urgently seeking the *necessary* and *just* global commitments—lie outside the limits of political "realism" and "reasonable" debate, it's time to ask who has the firmer grip on reality and reason.

And it's time to take the strongest nonviolent action. As climate-justice radicals, we need to be true to our understanding of the facts, and to our principles, our perception of right, even as conscience compels us to act—to be, crazy as it may sound, revolutionaries.

TWO

Prophets

We're called not to charity, or maybe even to justice—the scale of the injustice is so enormous it's hard to imagine ever rectifying it. What we're called to is something even more basic: solidarity.

. . . Our goal must be to make real the Gospel, with its injunction to love our neighbors. Not to drown them, not to sicken them, not to make it impossible for them to grow crops. But to love them.

—BILL MCKIBBEN, SERMON AT RIVERSIDE CHURCH, NEW YORK CITY,
APRIL 28, 2013

> *. . . Love the Lord.*
> *Love the world. Work for nothing.*
> *Take all that you have and be poor.*
> *Love somebody who does not deserve it.*
> *. . . As soon as the generals and the politicos*
> *can predict the motions of your mind,*
> *lose it. Leave it as a sign*
> *to mark the false trail, the way*
> *you didn't go. Be like the fox*
> *who makes more tracks than necessary,*
> *some in the wrong direction.*
> *Practice resurrection.*

—WENDELL BERRY, "MANIFESTO: THE MAD FARMER LIBERATION
FRONT," 1973

Naomi Klein—black-clad, sharp-tongued mistress of the global anticorporate left, friend to Occupiers and scourge of oil barons—stood outside a dressing room backstage at Boston's Orpheum Theatre one night in November 2012, a clear-eyed baby boy on her hip.

"I'm really trying not to play the Earth Mother card," Klein had told me the week before, as she talked about bringing Toma, her first and only child, into the world. But she needn't have worried about that image. She'd just finished fielding questions from a small gaggle of young reporters alongside 350.org's Bill McKibben, who had invited her to play a key role in the twenty-one-city "Do the Math" climate-movement road show that was arriving at the sold-out, 2,700-seat Orpheum that night (the original venue, Boston's Old South Church, had sold out in less than twenty-four hours, and the organizers had to scramble for a larger space). With a laugh, Klein noted to the reporters that McKibben's devastating *Rolling Stone* article, "Global Warming's Terrifying New Math"—which revealed, as we've seen, that the fossil-fuel industry had *five times* more carbon in its proven reserves than the science says can be burned if we're to have a shot at avoiding the worst, all of which it intends to extract and burn—had received no industry pushback.

"I mean, that's remarkable," she said, "for a piece like that, to not feel the need to correct the record in any way? *Actually, we don't plan to destroy the planet.*"

Then she offered an anecdote, as if to dispel any assumptions that she's a conventional green planet-saving type. Fresh from the Hurricane Sandy disaster zone, she described visiting an "amazing" community farm in Brooklyn's Red Hook that had been flooded. "They were doing everything right, when it comes to climate," she said. "Growing organic, localizing their food system, sequestering carbon, not using fossil-fuel inputs—all the good stuff." Then came Sandy. "They lose their entire fall harvest, and they're pretty sure their soil is now contaminated, because the water that flooded them was so polluted."

"So, yeah," she said, "it's important to build local alternatives, we have to do it, but unless we are really going after the *source* of the problem"—namely, she said, the fossil-fuel industry and its lock on Washington—"we are gonna get inundated."

For McKibben and Klein, going after that source meant, to begin with, going after the industry's very legitimacy. To that end, they and 350.org (on whose board Klein serves) were using the sold-out national tour to help launch a mostly student-led divestment campaign, modeled on the movement to divest from apartheid South Africa in the eighties, calling on universities—along with religious and philanthropic institutions, as well as state and local governments—to stop investing their endowments and pension funds in fossil fuels. As of that December, the effort had already spread to more than 150 campuses around the country—a number that would grow by mid-2014 to more than 400. By early 2015, twenty US schools had committed to divest, including Stanford University, which announced that it would divest from coal, with further divestment on the table. In September 2014, the Rockefeller Brothers Fund announced that the heirs to the fortune of John D. Rockefeller, founder of Standard Oil, would no longer invest in fossil fuels. The cities of Seattle, San Francisco, Portland, Oregon; Boulder, Colorado; Madison, Wisconsin; Ann Arbor, Michigan; Providence, Rhode Island, and many other municipalities, have voted for divestment. Many major faith communities, including the United Church of Christ (Congregationalist), the United Methodist Church USA, the Unitarian Universalist Association, numerous Quaker organizations, the Church of England, and the World Council of Churches have voted in favor of divestment. And the call for divestment has indeed spread globally, taking off in Europe, Australia, and New Zealand, with numerous educational and religious institutions making commitments.

The point of fossil-fuel divestment has never been the economic leverage (quite little) it might wield over some of the richest companies on Earth—or, for that matter, the potential economic benefit (quite real) of avoiding exposure to "stranded assets"—but instead a kind of moral leverage. Like Keystone, it has become a rallying point for a broad-based movement that would mount a resistance to what McKibben has called a "rogue" industry and its lobby. The moral logic of divestment says that this industry's profit model and political behavior are so reckless, it has forfeited its license to do business. Divestment, properly understood, is radical—and properly so. It points to the fact that the fossil-fuel industry as we know it must end—or civilization will end.

Later that night, on the Orpheum stage with McKibben, Klein told the audience: "Remember this moment. This was the moment we got *serious*."

To be sure, Bill McKibben and Naomi Klein had been plenty serious, in their respective ways, for a long time. Then fifty-two years old, McKibben, one of the world's leading environmental thinkers, writers, and activists, had been fighting the climate fight for more than a quarter of a century, ever since he wrote *The End of Nature*, published in 1989. In 2007 and 2008, together with a small band of students at Middlebury College, where he is the Schumann Distinguished Scholar, he cofounded 350.org, the global grassroots network that was the first serious effort—and a successful one—to spark a bottom-up movement to address climate change. (The name refers to 350 parts per million of carbon dioxide in the atmosphere, the amount that NASA climatologist James Hansen determined to be the likely maximum for a stable climate. We're now at around 400 and rising.) Today's grassroots climate movement would barely exist, and you wouldn't be reading this book, if it weren't for McKibben and the young cofounders of 350.org.

For her part, Klein "came of age politically," she told me, with the 1999 protests against the World Trade Organization in Seattle, when she was twenty-nine, shortly after which her international best-seller *No Logo* made her an intellectual star of the antiglobalization movement. Her 2007 book, *The Shock Doctrine: The Rise of Disaster Capitalism*, exposed the ways neoliberal free-market profiteers have exploited chaos and catastrophe in disaster zones, from hurricane-shocked New Orleans to "shock-and-awe"-shocked Iraq.

Seeing McKibben and Klein on stage together, a year after Occupy Wall Street, launching a mathematical and moral assault on the carbon oligarchy, said something significant about the direction the climate movement was taking. Or, rather, the direction McKibben and Klein argued that it *should* be taking, as they sought to merge climate and economic justice in a way that went beyond both traditional environmentalism and the old-school, climate-silent left. Each had a tough-love message for their own constituency: McKibben for an insular environmental movement that has been woefully ineffective on climate; Klein for a "siloed" left that has failed to grapple with the seriousness and urgency of the climate crisis. *Look*, they were saying, *this is it*: science tells us that time is running out, and everything you've ever fought for is on the line. Climate change has the ability to undo your historic victories and crush your present struggles. So it's time to come together, for real, and fight to preserve and extend what you care most about—which means engaging in the climate fight, really engaging, as if your life and your life's work, even life itself, depend on it. *Because they do.*

That November, Klein was hard at work on her book *This Changes Everything: Capitalism vs. the Climate* (2014), and we talked about what she was trying to communicate, and to whom. (We've since become colleagues writing for the *Nation*.) "The climate crisis," she said, "is the ultimate indictment of capitalism, certainly the model of capitalism that we have." But for too long, she said, "lefties thought climate was the one issue they didn't have to worry about, because big, rich green groups had it covered. And now it's like, actually, they really don't. That was a dangerous assumption to make." She argued that a serious response would require the left "to weave together disparate movements" under one banner—a "movement of movements" to meet this greatest of challenges. It would mean fighting to restore democracy and reinvigorate the public sphere; putting a stiff price on carbon to make polluters pay for wrecking the climate, while taxing wealth and bringing basic fairness into the system; reining in free trade and reregulating corporations; building alternatives to limitless profit and unsustainable growth while relocalizing our economies.

If there's a central idea driving Klein's work on climate, it's that the historic projects of economic and social justice and the urgency of climate justice are interdependent and inseparable—from the local level on up to the global. "Climate change," she said, "lends urgency to our fights for social justice like nothing else before." Klein's entry point into the climate issue was her interest in reparations for slavery and the historic crimes of colonialism. In 2008, covering the United Nations conference on racism known as Durban II, she realized that the reparations movement had shifted its focus to the idea of "climate debt"—that is, what the developed world, in tangible economic terms, owes to the people of developing nations who will bear (and are already bearing) the brunt of climate change, but have done little or nothing, historically, to cause it.

"The refusal to accept the importance of economic justice is the reason we have had no climate action. It's just that simple," Klein told me, essentially summarizing the position of the Global South at the UN table. "What has bogged down every round of UN negotiations on climate is the basic principle that the people who are most responsible for creating this crisis should take the lead and bear a heavier burden." And for poor nations, "there should be a right to develop a certain amount, to pull oneself out of poverty"—which means a transfer of wealth from North to South, to aid the fastest possible transition to renewable energy. The issue, of course, remains a sticking point for any global climate deal.

Klein first met McKibben and the 350.org team in late 2009 at the disastrous UN climate conference in Copenhagen, where she was pushing these issues. She was profoundly impressed, a friendship formed, and in April 2011 she joined 350's board.

"I'm sort of used to the environmental movement seeing me as a pain in the ass," she told me. "You know, when I talk about reparations and climate debt—it's seen as being off-message. You're just supposed to shut up about things like that. It's inconvenient—an inconvenient truth that can't be sold to the American public."

"So I was surprised when Bill invited me to be on the board, because I sort of thought that I was toxic," she said. "I think it just speaks to 350's deep understanding that these movements have to come together."

I asked her what she thinks it signifies to see Bill McKibben and Naomi Klein working together so closely.

"Climate change is the human-rights struggle of our time," she said. "It's too important to be left to the environmentalists alone."

And with that statement, whether intentionally or not, Naomi had just summed up not only the story of 350.org, and the climate movement it's done so much to build, but the story and the life's work of her friend Bill McKibben.

———

In October, a month before that show at the Orpheum, I drove up to Burlington, Vermont, on McKibben's and 350's home turf, to attend the "dress rehearsal" for the national "Do the Math" tour, set to launch in Seattle the day after the election. On assignment for *Grist* magazine, I'd been invited to spend some time "backstage" with the 350.org team, watch their run-throughs for the evening's production, and chat with McKibben and the others. And though I was on assignment, everyone knew (including of course my editor at *Grist*, Scott Rosenberg) that I was hardly there to cover it as a conventional reporter. Earlier that year I'd helped launch 350 Massachusetts, the independent grassroots network started by 350.org's allies at Better Future Project, a young nonprofit in Cambridge (on whose board I served until late 2014). And I was getting involved with other alumni in the nascent, student-led Divest Harvard campaign. You could say my assignment in Burlington was an inside job.

Perhaps this is where I should pause and explain that Bill McKibben and I have been acquainted for many years, having worked together occasionally when I was an editor at the *Atlantic* and the *Boston Globe*. As I dove deeper into the climate movement, we developed a kind of collaboration as fellow writers and activists. Not that Bill and I have ever really become close friends—we don't hang out with each other, we've never shared much about our personal lives—but we have a warm, collegial relationship.

I say all of this not only for the sake of transparency, but because Bill McKibben has had a significant impact on my own thinking about climate. That doesn't mean I'm incapable of stepping back and giving my honest view of his work. Much of what I write here he'll probably appreciate; some of it he may feel compelled to argue with. I don't know. But I'm sure I'll find out.

The Saturday night crowd on the Burlington campus was festive, raucous, pumped. When the man on the stage, Bill McKibben, said it was time to march not just on Washington but on the headquarters of fossil-fuel companies—"it's time to march on Dallas"—and asked those to stand who'd be willing to join in the fight, seemingly every person filling the University of Vermont's cavernous Ira Allen Chapel, some 800 souls, rose to their feet.

"The fossil-fuel industry has behaved so recklessly that they should lose their social license—their veneer of respectability," Bill told the audience. "You want to take away our planet and our future? We're going to take away your money and your good name."

Before heading up to Burlington, I'd asked Bill what the divestment campaign represented for the climate movement. How did it compare with the fight against Keystone XL, now more than a year since he and 1,252 others were arrested at the White House, leading Obama to delay his decision for further review?

"Fighting Keystone," he told me, "we learned we could stand up to the fossil-fuel industry. We demonstrated some moxie." But, he added: "We also figured out that we're not going to win just fighting one pipeline at a time. We have to keep all those battles going, but we also have to play some offense, go at the heart of the problem."

His "Do the Math" talk—which grew straight out of the *Rolling Stone* piece and the Carbon Tracker Initiative's analysis of fossil-fuel reserves—left no doubt about what that problem is, its scale, and its urgency. One simply cannot repeat this too often: To have any decent chance of preventing runaway warming within this century—to slow the process down and maybe, ultimately, stop it—something like 80 percent of fossil-fuel reserves must stay in the ground, forever, and the world must mobilize an all-out global shift to renewable energy.

Given the sheer amount of money at stake—tens of trillions of dollars—the odds of anything like that happening under current political conditions are roughly *nil*. Bill's point was that, if there's going to be any hope at all of preserving a livable climate, those political conditions must change decisively, starting now. And they can—but only if and when enough people, including those in power, understand the simple carbon math and realize that the fossil-fuel industry and its lobby are prepared to cook humanity off the planet unless somebody stops them.

The most affecting display in Burlington that night was a show of faces—people, all around the world, who since 2009 had organized and

New Englander, after all), he wrote with trademark but now seething understatement: "I'm a mild-mannered guy, a Methodist Sunday school teacher. Not quick to anger. So what I want to say is: This is fucked up. The time has come to get mad, and then to get busy."

Still every bit the soft-spoken, self-effacing speaker—and still droll, even laugh-out-loud funny, on stage—Bill had both darkened and toughened his message. It was as though, as a person of faith—yes, it's true, Bill McKibben is a lifelong churchgoer, Sunday school teacher, and sometime preacher—he had discovered his "prophetic voice." He may not thunder, that will never be his style, but he has become, I want to say, a sort of modern-day Jeremiah.

Bill flatly rejects any such comparison. "I'm not a prophet," he tells me. Full stop. But this much is undeniable: Bill seems to have remembered a basic truth of transformative social movements—that they're driven not by "positive messaging" (much less any simplistic, poll-tested "win-win" market optimism) but by deep moral conviction and moral outrage at intolerable injustice. The movements that change the world are moral struggles—and spiritual ones.

———

The fact that Bill is a lifelong churchgoing Christian is well known to his friends and colleagues, but no doubt strikes some of his secular readers and fans as strange, possibly a little embarrassing. Reporters have occasionally picked up on this aspect of his life, mentioned it in passing, but what's rarely if ever explored is just how central Bill's brand of faith is to his outlook and to the whole arc of his life's work. If you're one of those secular readers, I hope you'll bear with me here. This is not an exercise in self-righteousness, or evangelization, or whatever. Bill is not inclined to any of that (and neither am I). Nor is this by any means simple. No, what I'm trying to do is suggest, as best I can, who Bill McKibben is, where he's coming from, and what really drives him.

Perhaps I should start by mentioning that Bill McKibben was born in Palo Alto, California, in 1960, and grew up comfortably middle class in Toronto and in Lexington, Massachusetts, in the Boston suburbs; that

participated in 350.org's massive "global days of action," involving thousands of demonstrations in hundreds of countries, on every continent—and who were already suffering the impacts of climate change: in Kenya, Haiti, Brazil, India, Pakistan, the Pacific island nations, and many other places, including the United States. Projected on the big screen behind Bill, they were a profound reminder of the human costs of global warming. Likewise, Bill's message was about far more than math and carbon reserves. It was about justice and injustice, right and wrong—what you could call the moral equation.

On Saturday afternoon, after Bill and the rest of 350's small production crew ran through their script in the empty, echoing Ira Allen Chapel, I tagged along with them for lunch a short walk from the UVM campus. I asked Bill how the idea for the tour had been born.

It all went back, he said, to that seminal 2011 report from the Carbon Tracker Initiative in London. Bill told me that he and Naomi had both read that report early in 2012—and when they saw the numbers, they both realized the implications. "It exposed a real vulnerability of the fossil-fuel industry," Bill told me, "because it made clear what the outcome of this process was going to be if we continued."

There was a long pause, as he searched for just how to phrase what came next.

"There's always been this slight unreality to the whole climate-change thing," Bill went on. "Because most people, at some level, kept thinking—and rightly so—'Yeah, but no one will ever actually do this. No one will actually, knowingly, destroy the planet by climate change.' But once you've seen those numbers, it's clear, that's exactly what they're *knowingly* planning to do. So that changes the equation, you know?"

I noted that the people who built the fossil-fuel industry didn't set out to wreck the planet. It's an incredible accident of history that we ended up in this fix.

Bill nodded. There was, he said, "a sound historical reason" for the development of fossil fuels. "But that sound historical reason vanished the minute Jim Hansen basically explained, twenty-five years ago, that we're about to do in the earth. And now that we've melted the Arctic, it's well under way, at this point—it's outrageous, is all it is."

"Now we have the new, and in some ways, the most important set of facts since the original science around climate. This stuff on who owns what, in terms of reserves—it's the Keeling Curve of climate economics and politics," he said, referring to the graph of the ever-rising concentration of carbon dioxide in the atmosphere, one of the foundational discoveries of climate science. "These are the iconic numbers for understanding where we are now."

So could divestment generate enough leverage, economic or otherwise, to make a difference? I wanted Bill to explain how it was an effective strategy.

"I think it's a way to a get a fight started," Bill said without hesitation, "and to get people in important places talking actively about the culpability of the fossil-fuel industry for the trouble that we're in. And once that talk starts, I think it does start imposing a certain kind of economic pressure. Their high stock price is entirely justified by the thought that they're going to get all their reserves out of the ground. And I think we've already made an argument that it shouldn't be a legitimate thing to be doing."

And not just those existing reserves. Perhaps the most damning number to emerge out of the divestment fight is this: *674 billion.* That's how many dollars, according to Carbon Tracker, the top 200 publicly traded fossil-fuel companies spent in 2012 alone on exploration and development of *new reserves.* (It remains to be seen whether the recent collapse of oil prices will lead companies to pull back significantly on such spending.) In other words, in the face of global catastrophe, those who lead the industry have not only bankrolled a wildly successful effort to sow confusion and denial of climate science—and to obstruct any serious response to the crisis. In the meantime, they are busy digging us an ever deeper hole, committed to a business model that by any sane measure should be called genocidal.

Bill likes to say that what he and the rest of us are demanding is not radical—indeed, fighting to preserve the planet for our children and future generations is inherently conservative. The "real radicals," Bill will tell you, run fossil-fuel companies. Nothing could be more radical than

the catastrophic course they're pursuing: willingly changing the composition of the earth's atmosphere, consequences be damned.

At perhaps the key moment in Bill's "Do the Math" talk, he played a video clip of Exxon CEO Rex Tillerson at the Council on Foreign Relations in June 2012. Bill eviscerated the onscreen Rex in a darkly comic back-and-forth that would've made Jon Stewart proud. The Exxon chief, having made news by acknowledging that climate change is real and that warming "will have an impact"—while his company was spending, according to Bloomberg, as much as $37 billion per year exploring and drilling for more oil and gas—went on to express confidence that "we'll adapt." Agricultural production areas will be shifted northward, Tillerson suggests. (Never mind, Bill points out, that you can't just move Iowa to Siberia, and that there isn't any topsoil in the tundra.)

"It's an engineering problem, with engineering solutions," intoned Rex—who, as Bill noted on stage, was making $100,000 a day.

"No," Bill replied. "It's a greed problem. Yours."

In 2011 and 2012, the tone of the climate movement was shifting. Maybe it all went back to the failure of Copenhagen in 2009, the collapse of climate legislation in the Senate in 2010, and the disillusioning, infuriating lack of climate leadership by Barack Obama. With a kind of desperation, but with history as a guide, people began talking and writing in earnest about building a genuine grassroots movement, a peoples' movement, based on something more, something broader and deeper, than all the lobbying money and the corporate-style, K-Street-friendly communications strategies of the big green groups. A movement built on something more like moral outrage—and moral indictment.

Bill's tone had been changing as well. There were hints of it in those brutal opening chapters of his book *Eaarth,* released in the spring of 2010, where he surveyed the planet's damage and the all but certain ravages to come. And when the watered-down-to-nothing climate bill finally died in the Senate that summer, he let loose with a much-quoted broadside headlined "We're hot as hell and we're not going to take it anymore." As though finally venting emotions long suppressed (he's a

his father was a journalist who worked for *Business Week* and the *Boston Globe* and was arrested in 1971 on the Lexington town green supporting an antiwar protest by Vietnam veterans; that his family went to church on Sunday and the church youth group was a big part of his life; that he went to Harvard, where he was editor of the *Crimson*, and where he became good friends with the late great Reverend Peter J. Gomes, rector of Memorial Church; that he went straight on to the *New Yorker*, where he wrote "Talk of the Town" pieces for five years before quitting in protest when legendary editor William Shawn was forced out; that while in New York he and others started the homeless shelter at the famous Riverside Church; that he and his wife, Sue Halpern, a fellow writer, moved to the Adirondacks, and that he turned his full attention to what was happening to the planet; that they eventually moved to the Green Mountains of Vermont, overlooking the Champlain Valley, where he has taught at Middlebury College ever since. That they have a daughter named Sophie who's now in her early twenties. That Bill is seldom happier than when he's out in the woods after a snow.

But really, the first thing that should always be said about Bill McKibben is that he's the guy who wrote, while still in his twenties, *The End of Nature.* Not just the first book for a general audience about global warming, but without exaggeration, an American classic—a prescient tour de force, in which he reported on what was already the well-advanced science of human-caused climate change, and then proceeded to sketch the broader contours and substance of the subject as we still know it twenty-five years on. Rereading the book even now, you realize that there's been precious little new to say about climate change, in big-picture terms, since Bill explained it to us. Others had of course written about looming ecological catastrophe, "limits to growth," the Earth's carrying capacity, and so on. But when it comes to climate change, and its import, Bill was there first.

But he didn't just get the scoop; he went deep. Indeed, that was the real scoop. He thought hard about it, felt it, and wrote a bold, searching, moving—and, like most classics, in some ways idiosyncratic—extended essay on the meaning of what humanity has done to the earth and everything on it. Or more precisely, what our modern civilization has done to it. His subject is not only the fact that we've changed the composition of the atmosphere, but what it feels like as one struggles to comprehend the consequences, to take it all on board, philosophically and spiritually. I'm far from alone, I feel sure, when I say that the book has affirmed and clarified my sense of a spiritual crisis at the heart of the climate crisis.

How so? Consider that well before the term "Anthropocene" gained currency—the widely accepted idea among Earth scientists that we have left the Holocene and entered an Age of Man, in which humanity itself is now a geological force—Bill argued that our impact on the planet carries world- and worldview-altering significance. We've changed everything, even the weather. And so the idea of "nature" as something vastly larger and independent of humanity, of human society, cannot survive. We've delivered the death blow. Or as Bill writes:

> An idea, a relationship, can go extinct, just like an animal or a plant. The idea in this case is "nature," the separate and wild province, the world apart from man to which he adapted, under whose rules he was born and died. In the past, we spoiled and polluted parts of that nature, inflicted environmental "damage." . . . We never thought that we had wrecked nature. Deep down, we never really thought we could. . . .

Of course, as he acknowledges, "natural processes" go on. In fact, "rainfall and sunlight may become more important forces in our lives." The point is, he writes, "the meaning of the wind, the sun, the rain—of nature—has already changed." This realization leads him to what is perhaps the book's central statement: "By changing the weather, we make every spot on earth man made and artificial. We have deprived nature of its independence, and that is fatal to its meaning. Nature's independence *is* its meaning; without it there is nothing but us."

I want to come back to that last phrase—"nothing but us"—but before I do, it's crucial to understand the impact this realization has on Bill as a person of faith. Of course, in typical fashion, he disclaims: "I am no theologian; I am not even certain what I mean by God. (Perhaps some theologians join me in this difficulty.)" But he goes on to ask, "For those of us who have tended to locate God in nature—who, say, look upon spring as a sign of his existence and a clue to his meaning—what does it mean that we have destroyed the old spring and replaced it with a new one of our own devising?"

To answer that question, Bill finds himself drawn time and again to the Hebrew Bible's story of Job. First, however, he has to deal with the often heard environmental critique of the Bible's creation story, in Genesis, where God gives man "dominion" over the earth and commands him to "subdue" it. There in *The End of Nature*, Bill joins those who argue that this is far too narrow a reading, and observes that when we take the Bible as a whole, "the opposite messages resound." Many theologians, he rightly points out, "have contended that the Bible demands a careful 'stewardship' of the planet instead of a careless subjugation, that immediately after giving man dominion over the earth God instructed him to 'cultivate and keep it.'" But even this, he says, fails to really capture the depth of the Bible's ecological message. For that, he turns to Job—"one of the most far-reaching defenses ever written of wilderness, of nature free from the hand of man."

The Job story, of course, is a staple of Western literature, but to refresh, it goes like this: Job, we are told, is a wealthy, faithful, good, and just man, yet the devil makes a bet with God that if Job is stripped of all his possessions, his children, his happiness—really made to suffer—he will turn and curse God. The Lord is confident, and accepts the wager. Soon, Bill writes, "Job is living on a dunghill on the edge of town, his flesh a mass of oozing sores, his children dead, his flock scattered, his property gone." But Job, though he curses the day he was born, won't curse his Maker. He simply wants an explanation for his suffering. He maintains his innocence, and can't accept the orthodox view offered by his friends that he's being punished for some sin. Therefore God owes him an answer. *What have I done to deserve this?* Job demands.

Finally, God's voice speaks to him from out of a whirlwind, and the answer—as Bill puts it in his short book on Job, *The Comforting Whirlwind*—is "shockingly radical." It is God's longest soliloquy in the Bible,

and it is unsparing yet beautiful—perhaps, as Bill suggests, the foundation of Western nature writing. In Stephen Mitchell's striking, poetic
translation (the one Bill uses), God asks Job:

> Where were you when I planned the earth?
>   Tell me, if you are so wise.
> Do you know who took its dimensions,
>   measuring its length with a cord? . . .

> Were you there when I stopped the waters,
>   as they issued gushing from the womb?
> when I wrapped the ocean in clouds
>   and swaddled the sea in shadows?
> when I closed it in with barriers
>   and set its boundaries, saying,
> "Here you may come, but no farther;
>   here shall your proud waves break."

Bill has called this "God's taunt"—as if the Creator is saying, *You little
man, who do you think you are, demanding that I explain your suffering? Creation does not revolve around you.* God asks (again in Mitchell's translation):
"Who cuts a path for the thunderstorm / and carves a road for the rain— /
to water the desolate wasteland, / the land where no man lives; / to make
the wilderness blossom / and cover the desert with grass?" Indeed, that is
the rub. As Bill writes in *The End of Nature*: "God seems to be insisting that
we are not the center of the universe, that he is quite happy if it rains where
there are no people—that God is quite happy with the places where there
are no people, a radical departure from our most ingrained notions." To
Bill, this is a profoundly comforting thought—that we are subsumed into
something far larger, incomprehensibly powerful, and free of our touch.

And so back to Bill's question: What does it mean that we, or at
least some of us, have altered the atmosphere, changed the weather, the
storms—that, in effect, we are adding force to the whirlwind? When
God asks who set the boundaries of the oceans, Bill writes, "we can now
answer that it is us. Our actions will determine the level of the sea, and

change the course and destination of every drop of precipitation." There's a word for this, Bill likes to say: "blasphemy." We have usurped God.

And considering how our power-grab has worked out, that is not a happy way to be—whether you believe in the Bible's God or not. As Bill has said countless times in the past few years, we've taken creation's, the planet's, largest physical features—the Arctic, the oceans, the great glaciers—and we've broken them. We're perhaps a decade away from an ice-free Arctic summer. The oceans are now an ungodly *30 percent* more acidic, threatening the base of the marine food chain and all that depend on it. In other words, the end of nature is a pretty miserable place.

So it's not surprising that *The End of Nature* concludes on a dark and deeply pessimistic note. The book pulls no punches. It's too honest for that. What we've set in motion cannot be undone: "Now it is too late—not too late to ameliorate some of the changes and so perhaps to avoid the most gruesome of their consequences. But the scientists agree that we have already pumped enough gas into the air so that a significant rise in temperature and a subsequent shift in weather are inevitable." Even had the nations of the world begun "heroic efforts" in the 1980s, he writes, "it wouldn't have been enough to prevent terrible, terrible changes." We would still be committed, Bill informs us, to a warming far greater than humans have ever experienced.

That—in 1989. And he was right.

This leads him to say things like: "If industrial civilization is ending nature, it is not utter silliness to talk about ending—or, at least, trans-forming—industrial civilization." That would mean an acceptance of limits, an end to human hubris. Of course it sounds impossible—but what are the alternatives? "It could be that this idea of a humbler world, or some idea like it, is both radical and necessary, in the way that cutting off a leg can be both radical and necessary." He suggests that there are signs, however small, of such radical new thinking, as among the bio-centric "deep ecologists," citing Dave Foreman, founder of Earth First!, who drew inspiration from the writings of Edward Abbey (in particular his novel of eco-defense warriors, *The Monkey Wrench Gang*).

But that's about as much hope as Bill will allow himself at the end of the book. Remember that phrase: without nature as an independent force, "there is nothing but us." Ultimately, he is overwhelmed by a deep sadness and a sense of "loneliness." Tellingly, I think, in the book's final pages he even asks, "If nature has already ended, what are we fighting for?" He doesn't really have an answer. Not yet. Not in *The End of Nature.*

Of course, the *idea* of nature that Bill pronounced dead is itself a product of the human mind—an artifact of our particular evolution as a species, or really, of a particular civilization. And I want to say, it's as though Bill's crisis, the spiritual crisis of *The End of Nature*, is really the struggle to let go of his own conception—you might call it the biocentric, late-twentieth-century environmentalist conception—of what nature means. It's a struggle not unlike the struggle to let go of a deceased loved one.

And if that is the case—if in fact it is too late to save "nature," if there is "nothing but us"—then yes, the question Bill asks in the end demands an answer: What *are* we fighting for?

At this point, I want to propose another way of looking at Job—the way I've taken to viewing the story, one I've known since childhood, in light of our catastrophe and in light of my own deepest fear, and despair, for the future.

I see Job there on the waste, alone and naked in the dust, covered with ashes, tormented, diseased, his children dead—bereft of everything that he owned and loved. And I hear him crying out (Mitchell again):

> God damn the day I was born
>   and the night that forced me from the womb.
> On that day—let there be darkness;
>   let it never have been created;
>   let it sink back into the void. . . .
> On that night—let no child be born,
>   no mother cry out with joy. . . .
> Let its last stars be extinguished;
>   let it wait in terror for daylight;
>   let its dawn never arrive.

We are Job. Worse, our children are—alone on the ash heap, cursing the day they were born. Because, on our current course, Job is the vision of our future, our children's future—and for far, far too many, from the Philippines to the Rockaways, the vision of our present. It's not only that human beings have "ended nature" and usurped the place of God, not only that we have *inflicted* that death on nature, this catastrophe bearing down on us. We—and most of all the innocent, alive today and yet to be born—must *suffer* it. There is no comfort in the whirlwind.

And so, when I get to the end of *The End of Nature*, I see my friend Bill as a much younger man—a young man, alone, in the throes of the spiritual crisis of our time, who has yet to come to terms with the fact that what we are fighting for, now, is not only the earth but each other.

———

If *The End of Nature* is vulnerable to any major criticism, in my view, it's that the book as a whole doesn't have much of a social conscience. Which is odd, because the man who wrote it *does*. It's not that the book is blind to social and economic injustice—Bill points, for example, to the particular plight and dilemma of the global poor in the face of greenhouse emissions. ("The thought that people living in poverty . . . will curb their desire for a marginally better life simply because of something like the greenhouse effect is, of course, absurd. . . . They are going to moderate their 'lifestyles,' pare down their 'desires,' in order to avoid releasing carbon dioxide?") Nevertheless, the book's philosophical bent, and its lament for "nature" separate from humanity, plays easily into the caricature of a rich, elite environmentalism that cares more about "pristine" wilderness—real or imagined—than about people, about human suffering, and about human community.

But to leave it at that, to buy into the caricature, would be to miss what Bill McKibben is really about—and to unjustly paint him as simply another elite environmentalist. Because if there is one word that captures the bottom line of Bill's most important subsequent work—his books *Deep Economy: The Wealth of Communities and the Durable Future* (2007) and *Eaarth: Making a Life on a Tough New Planet* (2010)—it is that word "community."

And to understand how Bill arrives at that place, I believe you have to understand two things: his reverence for the Gospels, the teachings of

Jesus; and the influence of the writer and thinker he acknowledges as his intellectual mentor, the "finest writer at work in the English language," he calls him. I mean Wendell Berry, the Kentucky poet, essayist, novelist, farmer, and yes, Christian. "There is no one I admire more, as a writer and as a man," Bill has told me. "He really is a prophet."

Although Wendell Berry is often linked with the environmental movement—on issues from sustainable agriculture to the fight against mountaintop-removal coal mining in Appalachia—he transcends it and any narrow idea of what environmentalism means. For Berry it's as much about preserving the wildness in a handful of good topsoil—what he has dedicated his life as a small farmer to cultivating and preserving—as any wilderness. Above all, it's about community and love of neighbor, which means finding the right balance between human culture and the rest of creation.

"Creation" is a word that comes naturally to Berry. In fact, part of Berry's great importance for us now may be what you could call his crossover appeal: you don't have to be a secular-left environmentalist to love Wendell Berry. According to *Christianity Today*, some of his most avid readers in recent years have been young evangelicals. That should come as no surprise if you've been paying attention to the emergence of a vibrant evangelical environmental movement, and its underlying theology of "stewardship" and "creation care," over the past two decades. That movement produced the major statement of the 2006 Evangelical Climate Initiative—affirming that climate change is human caused, that it affects the poor most of all, and that Christians are called to take action—signed by more than two hundred prominent evangelical leaders. And it has given rise to national groups like the Evangelical Environmental Network, New Evangelical Partnership for the Common Good, Young Evangelicals for Climate Action, and others. But Berry, who doesn't necessarily fit easily within evangelicalism, was making the biblical case for "stewardship" long before it became a buzzword in Christian circles.

While Berry is a serious Christian—or, as he has said, "a person who takes the Gospel seriously"—he is also a forthright critic of modern Christianity's complicity in our ecological crises. "The certified Christian seems just as likely as anyone else to join the military-industrial conspiracy to murder Creation," he writes in "Christianity and the Survival of Creation," his 1992 lecture at the Southern Baptist Theological Seminary in Louisville. And yet, he argues in that same lecture, despite the "catastrophic discrepancies" he sees between the Bible's teachings and "allegedly respectable Christian behavior," he cannot simply dismiss Christianity—its potential contributions are too great. And so Berry holds out the "better possibility" that for those like himself—"whose native religion, for better or worse, is Christianity . . . an intimate belonging of our being"—the Christian faith "should survive and renew itself so that it may become as largely and truly instructive as we need it to be." Considering the influence of Christianity in the United States and globally, much is at stake. "On such a survival and renewal of the Christian religion," Berry writes, "may depend the survival of the Creation that is its subject."

What would that kind of survival and renewal look like? It's helpful to revisit a prophetic essay Berry wrote more than a decade before that Louisville lecture. In "The Gift of Good Land," published in 1979, Berry sets out to make "a Biblical argument for ecological and agricultural responsibility," and in the process he develops a scriptural and moral connection between *land* and *community* and *justice*—a connection that, I believe, has been profoundly important to Bill McKibben, as a writer and an activist.

The "good land" to which Berry refers is none other than the Promised Land—"a divine gift to a *fallen* people." And what the people are given, he notes, "is not ownership, but a sort of tenancy." The land is not a "reward," Berry writes. "It is made clear that the people chosen for this gift do not deserve it," and that "having failed to deserve it beforehand, they must prove worthy of it afterwards."

The first condition the land's tenants must meet is to be "faithful, grateful, and humble; they must remember that the land is a gift." But it's the second condition, as Berry spells it out, that really strikes home: "They must be neighborly. They must be just, kind to one another, generous to strangers, honest in trading, etc." It's worth quoting Berry here at some length:

These are social virtues, but, as they invariably do, they have ecological and agricultural implications. For the land is described as
an 'inheritance'; *the community is understood to exist not just in
space, but also in time.* One lives in the neighborhood, not just
of those who now live "next door," but of the dead who have
bequeathed the land to the living, and of the unborn to whom
the living will in turn bequeath it. But we can have no direct
behavioral connection to those who are not yet alive. *The only
neighborly thing we can do for them is to preserve their inheritance*:
we must take care, among other things, of the land, which is never
a possession, but an inheritance to the living, as it will be to the
unborn. (emphasis added)

The third condition, then, as Berry writes, is to take care of the land,
"to practice good husbandry." The scripture offers, he argues, "a perfect
paradigm of ecological and agricultural discipline, in which the idea of
inheritance is paramount." What the land produces, you may use, but
above all "you must preserve the fertility of the fields."

What Berry is pointing to here is nothing less than an ethical source,
a deep spiritual wellspring, of *environmental* and *social* and *intergenerational justice.* And it's fair to say that the essay's conclusions land with
prophetic force: "It is a contradiction," he writes, "to love your neighbor
and despise the great inheritance on which his life depends." He goes
on, turning to address members of his own faith, who have forgotten the
conditions by which they are bound:

If "the earth is the Lord's" and we are His stewards, then obviously some livelihoods are "right" and some are not. Is there, for
instance, any such thing as a Christian strip mine? A Christian
atomic bomb? . . . Is it Christian to profit or otherwise benefit
from violence? Is there not, in Christian ethics, an implied re-
quirement of practical separation from a destructive or wasteful
economy?

At the end of the essay, he offers a statement as radical as any prophet's:

It is possible—as our experience in *this* good land shows—to exile ourselves from Creation, and to ally ourselves with the principle of destruction. . . . If we are willing to pollute the air—to harm the elegant creature known as the atmosphere—by that token we are willing to harm all creatures that breathe, ourselves and our children among them. There is no begging off or "trading off." You cannot affirm the power plant and condemn the smokestack, or affirm the smoke and condemn the cough.

It is impossible to "live harmlessly," Berry knows. But when we "ignorantly, greedily" destroy our inheritance, which is also that of our children and future generations, it is not just a sin; it is a desecration. "In such desecration," Berry writes, in the essay's final words, "we condemn ourselves to spiritual and moral loneliness, and others to want."

We condemn others to want. Berry couldn't be clearer: our relationship to the earth, the land, is the basis of our relationship to the community, to each other.

I've spoken in terms of prophets and prophetic voices, even invoking earlier the Hebrew prophet Jeremiah. I chose the latter with intention. American preachers and social critics from the seventeenth century on have often imitated the biblical Jeremiah and the particular form of sermon, developed by the Puritans, that bears his name. The jeremiad, as we call it, is at once an indictment of and a lament for the community's sins, as well as an exhortation to return to the true faith, or the founding principles.

It is no stretch to place Wendell Berry in this American tradition—a tradition that links him to the radical Christian currents of abolitionism, the Social Gospel, and the sermons and speeches of Martin Luther King Jr. And seen in this way, it's clear that for Berry what we need to restore is not only our right relationship to the earth—to repair our broken covenant with creation and Creator—but our relationship to one another, our human community. In the Berrian jeremiad, what we have to get back to, in the deepest sense, *is* community. And as Berry shows us, there can be no true community without love of neighbor—that is, without justice, social and generational, in place and in time.

"Corporate industrialism," Berry writes in "It All Turns on Affection," his 2012 Jefferson Lecture at the National Endowment for the Humanities, "has failed to sustain the health and stability of human society. Among its characteristic signs are destroyed communities, neighborhoods, families, small businesses, and small farms. It has failed just as conspicuously and more dangerously to conserve the wealth and health of nature." In the end, he says, "the land and the people . . . have suffered together, as invariably they must. Under the rule of industrial economics, the land, our country, has been pillaged for the enrichment, supposedly, of those humans who have claimed the right to own or exploit it without limit." And yet, he concludes, none of this was necessary or inevitable. There has always been, as there is now, a choice.

"We do not have to live," Wendell Berry reminds us, "as if we are alone."

———

"Watching all of this play out, in the real world, is a very unsettling thing. To have known in advance what was going to happen, and then to watch it happen, is extraordinarily hard."

Bill was sitting across from me at a small table in Carol's Hungry Mind Cafe in Middlebury. It was December 14, 2013, and zero degrees outside—a mass of Arctic air over New England and a major snowstorm sweeping down from Canada.

"In some ways, it's easier for me than for other people, because I got through my own angst about all this relatively early on. The first few years after writing *The End of Nature* were very difficult—dark."

Bill's voice lowered, almost to a whisper, as it tends to do when he's answering difficult questions, personal questions, and he spoke slowly, with long pauses, not haltingly but deliberately. He rested his elbows on the table, and for a moment held his head in his hands.

"But I've had plenty of time to—in some ways, I've come to terms with it. On the other hand, it's very unsettling to watch it play out, and realize one was unable to prevent it. It's the equivalent of those bad dreams where you're trying to warn people about something and you can't get anyone's attention, and your voice doesn't work."

He looked up. "I mean, I wish I'd figured out sooner what was going on, and started things like 350 long before. We needed a mass movement a long time ago."

Another long pause.

"So, yeah, there's a certain amount of anguish."

We'd been talking about the book of Job, and about Wendell Berry, and I was asking Bill about that sense of spiritual crisis at the heart of all this. I wanted to know what led him out of the cosmic loneliness of *The End of Nature* to the people-centered commitment that has driven him as an activist.

God's response to Job is "very radical," Bill said—"you've gotten this wrong, you're not the center of the world." Likewise, he said, "I think Wendell Berry is very radical. He says, you've gotten this wrong—individualism is not the key, community is. And I've taken that message—that's been very important to my understanding."

For Bill, that primacy of community connects back to his and Berry's shared religious faith. "This comes straight out of my own spiritual tradition," he said. "One is clearly, in the Gospels, commanded to love one's neighbor—something we're not doing a very good job of—and in the Hebrew Bible, to pay some attention to stewardship, to the future. Something we're clearly doing an extremely poor job of. We're the opposite of stewards."

I reminded him of Berry's Louisville lecture, "Christianity and the Survival of Creation," and wondered how something like that resonates with him.

"Christianity is much too important a resource, for the reasons that Berry outlines, to dismiss," Bill said. "Though one is tempted to over and over again, because so little of what passes for organized Christianity seems to bear much resemblance to the very clear call of the Gospel." He added: "But one has to remember that very little in one's own life bears very much resemblance to it either."

So what was it, I wondered, that really moved him to engage—to go all-in as an activist, to commit his life to movement building. When he thinks about it, Bill said, he sees the process in stages, the first of which was simply recognizing the true dimensions of the crisis. That's largely what *The End of Nature* was about. But it didn't stop there.

"Realizing the grave dimensions of this, in terms of peoples' lives," Bill told me, was the next step. "Remember that in 1989 it was still an abstract threat. You couldn't really find people whose lives were being turned upside down by what was unmistakably climate change. And so for me, the argument at first was more philosophical than anything else. But by the turn of the millennium or so, that was no longer true. I was in places like Bangladesh, getting dengue fever, and really sensing it on a more visceral level."

The final stage, he said, was grappling with why nothing was being done, at the social and political level, despite the obvious threat. "None of this should actually have to be happening," Bill said emphatically. "Viewed rationally, we should have just gotten to work. We knew a quarter-century ago what the problem was; we knew more or less, basically, how to fix it—and it was a series of quite rational changes in policy and technology that were endorsed by every economist that I can think of. And yet nothing happened. But some part of me, child of the Enlightenment and the Ivy League that I am, thought that our institutions would at least make a good faith attempt to do something about it."

So the real turning point, Bill says, "was understanding that the *reason* nothing was being done was because there was someone on the other side of this. That it wasn't an abstract battle—there were certain human beings completely committed to making sure it happened"—that is, the fossil-fuel lobby, intent on continuing business as usual, whatever the consequences. "Out of all those things was born 350.org."

If there's a particular moment that Bill keeps returning to, in his talks and his writing, which captures his personal turning point, it's that experience in Bangladesh. As he describes it in *Eaarth*, it was the summer of 2000, and he was doing research in Dhaka. He knew the country faces some of the worst of what climate change has in store—that the Brahmaputra and Ganges Rivers will dry up as the Himalayan glaciers that feed them slowly disappear, that sea-level rise is already pushing saltwater from the Bay of Bengal into the agricultural heartland. But that summer, he writes, "everyone was focused on a much smaller result of global warming"—a species of mosquito, *Aedes aegypti*, that was causing the first major outbreak of dengue fever in Dhaka.

"I was spending a lot of time in the slums of Dhaka," Bill writes, "and I got bit by the wrong mosquito." Sicker than he'd ever been—"the kind of fever where sweat runs off your outstretched arm like rain off a gutter"—he nevertheless had it easy. "I was healthy going in and young and well fed, and so I didn't die. Plenty of people did: children, old people, weak people."

As we saw in *The End of Nature*, Bill has always been conscious of the particular plight of the global poor in the face of climate change, but now, he makes clear, the situation is far more extreme. "Global warming," he goes on to write there in *Eaarth*, "turns the idea of 'development' into a cruel joke." And the fact that the world's poor have done nothing to create the climate catastrophe has long been one of the major obstacles to serious action. "In that world," he asks in *Eaarth*, "how do you sit down and negotiate a global climate pact?" He points to the work of Tom Athanasiou and Paul Baer, directors of the Greenhouse Development Rights Network, who note that developing nations not only have justice on their side, but physics and economics: "After all, if they burn all their coal," Bill writes, "there's nothing the rest of us can do to ward off global warming."

This is what the global climate talks have really been about, Bill points out, ever since Rio de Janeiro in 1992. If the wealthy and historically responsible nations of the North will agree to "share" the remaining atmosphere with the South—rather than devour what remains of the atmospheric carbon budget—then the South will develop on a different path. In the simplest terms, Bill writes, "the deal goes like this: You give us enough windmills, and we won't burn our coal. You rebuild our factories so they're efficient, and we won't burn our coal. You come up with some other attractive ways to lift us out of poverty, and we won't burn our coal." For twenty years, Bill notes, everyone has known this would have to be the bargain, and for twenty years everyone has gamed the talks. The result, now, is that we've run out of time, and the odds of a global agreement in Paris with the force to head off catastrophe are vanishingly small.

In 2007's *Deep Economy*, Bill keeps returning to the problems of development, drawing especially on his reporting among factory workers in China—for whom coal means warmth in the winter, and producing cheap consumer goods means a chance at climbing out of poverty. At the same time, he constructs a compelling, erudite argument for economics at a smaller, local, human and workable scale, what he calls an "economics of neighborliness." It's a case for community, neighborliness, at both the local and the global level—not only as remedies for our environmental problems, but quite possibly our only hope of building the resilience to get through what's coming. In this "changed world," he writes, "if you're a functioning part of a community that can meet at least some of its needs—for food, for energy, for companionship, for entertainment, for succor—then you're more secure." This will require economies based not on maximum short-term growth, but maximum long-term durability. "The key questions will change from whether the economy produces an ever larger pile of stuff to whether it builds or undermines community—for community, it turns out, is the key to physical survival in our environmental predicament and also to human satisfaction."

And yet Bill knows full well that the sheer "momentum of physics," as he puts it, is bearing down on us, making the vision of any ideal economy seem more and more distant, if not utopian—while his vision of community as our best hope for resilience, even survival, becomes all the more urgent. This was clear when I brought up Wendell Berry's argument for community and rootedness in one's place. Berry, Bill said, "has argued very strongly for the primacy of staying in one's community and doing the work of building one's community so that it works. And I think that's exactly right. Absent the physics of climate change, that's all I'd want to be doing."

But the physics of climate change, Bill said, "is a wild card thrown into all of this. In order to be able to build strong local communities and so on, we also have to deal with this overarching problem. Because as we learned in Vermont, with Hurricane Irene, it doesn't matter how wonderful your organic farm is—if it's under water, you're still screwed."

"Left to its own devices," Bill said, "I do think the world wants to move toward smaller community, toward working scale. I think we're coming to understand, in psychology and economics, lots of disciplines, that there's a great deal to be said for community and connection. Human connection is what we've sacrificed as we built consumer capitalism, and we could get a lot of it back."

The problem, he says, is that this kind of shift takes a long time, generations, and we're never going to get there "intact" unless we deal with the pace of global warming, which already looks to be getting out of hand. "My fear," he told me, "is that by later in the century we're just having to divert all our resources and all our attention to a kind of ongoing emergency response project, until it gets too big and we can't do it anymore. In my dark moments, that's what the future looks like."

I asked if he finds it hard to be honest with people, maybe even with himself, about the situation we're now in.

"For many years, I have been very insistent on saying to people that stopping global warming is no longer one of the options. We can keep it from getting worse than it would otherwise get—and the difference between a world where the temperature has gone up two degrees, and a world where it's gone up four degrees, is a difference very much worth fighting for. It might prove to be the difference between a world with civilization and a world without it."

Still, he said, the climate fight, because of the nonrepealable laws of physics, is different from most other moral struggles. It's not just the fact that there's a deadline—an unknown point at which things spin out of control—but that it's already too late to prevent dire consequences. There's no foreseeable point at which victory is attained—there's simply too much carbon already in the atmosphere.

"You can't really, plausibly, give an 'I have a dream' speech for climate change," Bill said, "because the two possibilities are a miserable century and an impossible one."

And so our job now, he said, "is to prod and push the system as hard as we can—to force the spring. To speed up, catalyze, the response—make it happen more quickly. And the good thing about catalysts is that relatively small quantities of things can change the chemistry pretty quickly sometimes."

"Look," he said, "we just happen, unluckily, to be alive at one of—if not *the*—hinge moments in human history. One has to be wary of saying this, because everyone thinks they live in the most special of times. But I'm afraid that the physical evidence around us is pretty clear—that we're seeing phenomena like the melt of the Arctic and the acidification of the oceans that haven't been seen for tens of millions of years."

"We're at one of those very deep moments," he said, "where we're going to make some choices, or not make some choices, that are very fateful, all-or-nothing choices. We're in a very exposed place, morally and psychologically."

I asked if he dwells much on his daughter's life and her future.

"Sophie—she'll be twenty-one this spring. Of course, I think about it all the time, as all parents do at some level. But I try not, at this point—I don't find it useful, day to day, to spend a lot of time worrying about whether we're winning or losing. It seems sufficient just to get up and go to work, and see what can be done."

Do you believe in God? I asked him.

He thought for a moment. "Yeah, I mean, I have more belief in the Holy Spirit, I think. Of the Trinity, that's the part that's most apparent to me."

If you believe in any one of the three, I said, then don't you believe in them all?

"You're a better theologian than me," he said, smiling, and we laughed.

The laughter felt good. We needed it.

He thought for a moment again, looking down at the table, then up at me.

"I see around me all the time—when I look at those unlikely sixty or seventy thousand photos in the 350 Flickr account, of people in every corner of the world engaged in this movement—yeah, clearly, there's a force at work in the world, beyond sheer self-interest, that is moving people to act."

THREE

Organizing for Survival

Already, climate change is shaping up to be as unfair as disasters have ever been. . . . It too is a democracy question, about who benefits, who loses, who should decide, and who does. Surviving and maybe even turning back the tide of this pervasive ongoing disaster will require more ability to improvise together, stronger societies, more confidence in each other. It will require a world in which we are each other's wealth and have each other's trust.

—REBECCA SOLNIT, *A Paradise Built in Hell: The Extraordinary Communities That Arise in Disaster,* 2009

Even if segregation is gone, we will still need to be free; we will still have to see that everyone has a job. Even if we can all vote, but if people are still hungry, we will not be free. . . . Remember, we are not fighting for the freedom of the Negro alone, but for the freedom of the human spirit, a larger freedom that encompasses all mankind.

—ELLA J. BAKER, SPEECH IN HATTIESBURG, MISSISSIPPI, JANUARY 21, 1964 (QUOTED IN HOWARD ZINN, *SNCC: The New Abolitionists*)

Hilton Kelley stood smiling in the clear April sunshine outside Kelley's Kitchen in Port Arthur, Texas, his beloved hometown, and extended his hand. A big-framed man, with generous, gentle eyes and white stubble, Kelley was fifty-three years old when I met him that day in the spring of 2014.

The sign on the small corner restaurant read Delicious Home-Cooked Food, but Kelley's Kitchen was no longer serving. Kelley had opened it up in 2010, and managed to keep it running for about two and a half years. "It was going fairly well," he told me. "But, you know, the town really doesn't get a lot of foot traffic on this side of Port Arthur anymore."

Kelley's Kitchen was the only structure left standing on its block of Austin Avenue, just two blocks from Procter Street, the Gulf Coast city's main downtown thoroughfare. In every direction were more vacant lots and dilapidated buildings—windows blown out, many of them empty for years, even decades. In the bright sun, the streets at midday on a Friday were ghostly quiet.

"This area was once a thriving community," Kelley said. "It was traffic up and down Austin Avenue here."

Kelley invited me inside, out of the glare, and we sat at one of the tables in the well-kept place, which he still rented out for private parties and special occasions—there was even a small dance floor complete with a shiny disco ball. But that's not all that went on at Kelley's Kitchen. The space doubled as the office of the Community In-Power & Development Association, or CIDA—the small, tough, grassroots community advocacy and environmental-justice organization that Kelley founded in 2000, soon after returning to Port Arthur from California, where he was working in the movie industry as an actor and stuntman. In 2011, he received the prestigious Goldman Prize for his environmental-justice activism. Kelley has testified before the Texas Legislature and the US Senate, addressed UNESCO in Paris, and met President Obama at the White House.

Just a few blocks from where we sat is the historic African American community of West Port Arthur, where Kelley was born and raised in the Carver Terrace housing project, on the fence line of two massive oil refineries—one owned by Valero (formerly Gulf Oil) and the other by Motiva (formerly Texaco). In fact, the recently completed expansion of the Motiva refinery, which Kelley's group fought hard against, made it the largest in the nation, having more than doubled its capacity to six hundred thousand barrels of crude per day. Nearby are five more petrochemical plants and the Veolia incinerator facility. Port Arthur is also on the receiving end of the Keystone XL tar-sands pipeline, the southern leg of which—cutting through East Texas communities—went operational in January 2014. But the industry had brought few jobs to West Port Arthur, where unemployment was over 15 percent. Many workers commute to the plants, and economic development has moved north since the eighties, along with white flight, to the newer Mid-County area along Highway 69 toward Nederland, where you'll find a sudden explosion of malls, big-box stores, hotels, and theme restaurants with busy parking lots.

And yet the economic abandonment of the downtown area and West Port Arthur, in the very shadow of one of the world's most profitable industries, isn't even the whole story—there's also the pollution, some of the most toxic in the country. "One in five West Port Arthur households has someone in it with a respiratory illness," Kelley said. "One in five." The county's cancer mortality rate is 25 percent higher than the state average. Toxic "events"—whether from gas flares or accidents—are common, emissions often darkening the sky, fumes wafting into the neighborhood. The community is downwind of several of the refineries nearby. "If one isn't flaring or smoking, another one is," Kelley said. "At least twice a month, we're going to get some flaring and smoke from one of them." As much as he can, he documents the events with photos and video. "Sometimes it'll be really pungent," he told me, "to the point where it stings the nose and eyes." But apart from these incidents, he added, there's the constant day-to-day toxic menace in the air. "It's not always what you see—it's what you don't see. A lot of these gases are very dangerous. Sometimes newcomers will smell it and we can't, because we're desensitized to it."

Kelley had offered to show me around and give me the fence-line tour on the west side, the community where he grew up. I knew about his accomplishments with CIDA—among other things, how they'd successfully pressured both Motiva and Valero, the former to install state-of-the-art equipment to reduce toxic emissions and pay for a community development center, and the latter to fund a new health clinic. And I understood that CIDA is more than an environmental-justice group: its mission is to educate, empower and revitalize the community, working especially with young people. I knew that Kelley has made a real difference since returning home.

But before we left Kelley's Kitchen, I needed to ask him about another threat—one that, given Port Arthur's economic and racial marginalization, its proximity to dangerous petrochemical infrastructure, and its location on the gulf, could ultimately be the most devastating of all.

Yes, he answered, "we are seeing some of the impacts of climate change around here, as a matter of fact." The rising sea level has washed out parts of Highway 87 between Port Arthur and Galveston. "They've abandoned the road," Kelley said. And the ferocity of hurricanes, from Katrina and Rita to Ike, has shaken even Port Arthur natives like him. They were spared the worst of Katrina in 2005, "but Rita came very soon after that, and that's when we got hit hard," Kelley said. "I mean, a lot of the houses are gone. You can still see the FEMA tarps on some of the roofs today. A lot of homes that were once inhabited are now abandoned, because the federal dollars didn't come in soon enough and the houses just dry-rotted." The residents of Port Arthur haven't faced the kind of epic flooding that was seen in New Orleans, but with Hurricane Ike they came close. "Ike brought in a huge surge, and it reached right to the top of our hundred-year levee but didn't breach it." Even so, the roof of Kelley's old office was torn off: "The rain just poured in and destroyed everything."

I'd heard about Port Arthur, well known to environmental-justice advocates as one of the country's most egregious "sacrifice zones." But nothing prepared me for the physical reality of the place—a decaying, all-but-forgotten urban landscape inhabited by a struggling and precariously resilient community. As you drive west and north out of downtown, the refineries stretch for miles, at times towering over you like something out of dystopian science fiction. And yet this is not some futuristic scenario—it's here and now. And those same smokestacks that are poisoning the inhabitants of Port Arthur are part of a global fossil-fuel infrastructure that has trapped us in its political-economic grip, threatening civilization and the future of life on Earth—threatening not only the children of Port Arthur but everyone's children, everywhere, including my own.

But here's the thing: if you live in West Port Arthur and toxic emissions have ruined your health, or your child can't go to school because she can't breathe, or you can't find a job and feed your kids and see no way out of the projects—or all of the above—then you're probably not thinking about some future catastrophe. You're living in one. And what's true of Port Arthur is true of communities across the Gulf Coast and across the continent—and the world.

———

The struggle for climate justice is a struggle at the crossroads of historic and present injustices and a looming disaster that will prove to be, if allowed to unfold unchecked, the mother of all injustices. Because the disaster that is unfolding now will not only compound the suffering of those already oppressed (indeed, is already compounding it); it may very well foreclose any hope of economic stability and social justice for current and future generations.

Why, then, does the term "climate justice" barely register in the American conversation about climate change? Lurking in that question is a tension at the heart of the climate struggle: a tension between the "mainstream" climate movement (dominated by largely white, well-funded, and Washington-focused green NGOs) and those—most often people of color—who have been fighting for social and environmental justice for decades.

Nobody has worked longer and harder at this intersection of climate and environmental justice than Robert D. Bullard, the celebrated sociologist and activist, author of eighteen books, who is often called the father of the environmental-justice movement. In 1994, he founded the Environmental Justice Resource Center, the first of its kind, at Clark Atlanta University, and since 2011 he's been the dean of the Barbara Jordan–Mickey Leland School of Public Affairs at Texas Southern University (TSU) in Houston. It was Bullard who introduced me to Hilton Kelley, and I knew he could offer insight into the relationship between the environmental-justice and climate movements.

"Climate change looms as *the* global environmental-justice issue of the twenty-first century," Bullard writes in 2012's *The Wrong Complexion for Protection: How the Government Response to Disaster Endangers African American Communities*, coauthored with his longtime collaborator Beverly Wright, founding director of the Deep South Center for Environmental Justice at Dillard University in New Orleans. "It poses special environmental justice challenges for communities that are already overburdened with air pollution, poverty, and environmentally related illnesses." Climate change, as Bullard and Wright show, exacerbates existing inequities. "The most vulnerable populations will suffer the earliest and most damaging setbacks," they write, "even though they have contributed the least to the problem of global warming." As if to prove the point, their book project was delayed for more than two years by Hurricane Katrina, which destroyed the Deep South Center's computer files and devastated Wright's New Orleans East community. Her chapters documenting the unequal treatment of the city's African Americans in the Katrina recovery are essential reading.

Bullard's landmark 1990 book *Dumping in Dixie: Race, Class, and Environmental Quality* established the empirical and theoretical basis—and, for that matter, the moral basis—of environmental justice. Through his early work, beginning in 1978, on the siting of urban landfills in Houston's African American neighborhoods—as well as the siting around the country of toxic waste and incineration facilities, petrochemical plants and refineries, polluting power plants, and other industrial facilities—Bullard has systematically exposed the structural and at times blatant racism, which he names "environmental racism," underlying the disproportionate burden of pollution on communities of color, especially African African communities in the South. His work has done much to set the agenda of the environmental-justice movement.

In 2014, the movement was marking the twentieth anniversary of Executive Order 12898, signed by President Bill Clinton in February 1994, which explicitly established environmental justice in minority and low-income populations as a principle of federal policy. That year also marked the fiftieth anniversary of the Civil Rights Act—a fitting coincidence, as Bullard liked to point out, because the "EJ" executive order reinforced the historic 1964 law. Nevertheless, as Bullard and his TSU colleagues wrote in a report titled *Environmental Justice Milestones and Accomplish-*

ments: 1964 to 2014, "The EJ Executive Order after twenty years and three U.S. presidents has never been fully implemented." That would qualify as an understatement.

I sat down with Bob Bullard that April in his office at TSU, where we had two lively and substantive conversations. I'd interviewed him once before, the previous August, and in the meantime he'd been much in demand. In September, he had received the Sierra Club's John Muir Award, its highest honor (and the club went on to name its new environmental-justice award after Bullard); in March, he had delivered the opening keynote address at the National Association of Environmental Law Societies conference at Harvard Law School, assessing environmental justice after twenty years (former EPA chief Lisa Jackson was the other keynoter). When we sat down together in Houston, I'd seen him just a few days earlier in Cambridge, where he received two standing ovations from the jam-packed Harvard audience.

Bullard, who grew up in small-town Alabama, speaks with an orator's cadences and a comedian's timing. At sixty-seven, he had a fighter's glint in his eye and an irresistibly mischievous grin above a Du Boisian goatee (he calls W. E. B. Du Bois his intellectual hero). In Houston, I asked him about the relationship between environmental justice, traditionally understood, and climate justice.

Bullard likes to start with a history lesson. In 1991, he helped convene the First National People of Color Environmental Leadership Summit in Washington, DC, where seventeen "Principles of Environmental Justice" were adopted. At the 1992 Earth Summit in Rio, those principles were circulated in several languages, but it wasn't until 2000, in The Hague, that Bullard joined other leaders and groups from around the world for the first "climate justice summit," which met in parallel with the sixth United Nations climate conference, or COP6.

"It was a very transformative time," Bullard recalled. "When environmental-justice groups and groups working on climate, on human rights and social justice and civil rights, came together in The Hague in 2000, 'climate justice' was not a term that was universally used." At that summit, he told me, "we said that climate justice has to be the centerpiece in dealing with climate change. If you look at the communities that are impacted first, worst, and longest—whether in Asia, Africa, Latin America, or here in the US—when you talk about the majority of people around the world, climate justice is not a footnote. It is the centerpiece." Globally, he points out, climate justice "is not a minority view, it's the majority view."

Here in the United States, Bullard said, "equity and justice get a footnote." In terms of framing the climate conversation, he said, it's been a struggle to make sure that justice is given parity with the science. "That's the rub," Bullard told me. "And that's why the climate movement has not been able to get traction like you'd think it would, given the facts that are there. The people on the ground who could actually form the face of climate change, be the poster child of global warming—they're almost relegated to the fringes. And that is a mistake." In the United States as well as globally, Bullard said, "we know the faces, we know what they look like. We know the frontline communities, the frontline nations. But to what extent do we have leadership that's reflective of communities that are hardest hit? Very little has changed over the last twenty years when it comes to who's out there."

This criticism of mainstream climate and environmental groups, and the foundations who fund them, has been leveled countless times—and it has stuck for a reason. Until very recently—witness the widely noted diversity of the massive September 2014 People's Climate March in New York City, organized by a broad coalition that included environmental-justice and labor organizations—frontline communities, and especially communities of color, have been conspicuously underrepresented in the climate movement.

And yet, I observed, even with such inherent tensions, climate justice ought to be the most unifying concept on the planet—if only for the simple reason that people everywhere tend to care about their children and grandchildren. I had asked Bullard earlier about the idea of *intergenerational* justice—based on the fact that, along with those in the poorest and most vulnerable communities around the world, today's young people and future generations will bear vastly disproportionate impacts of climate change. Isn't climate justice really environmental justice writ large—on a global scale—yet with this added generational dimension?

"Exactly," Bullard said. "And for me, that's the glue and the organizing catalyst that can bring people together across racial and class lines."

In that case, I wondered aloud, if the central purpose of the climate movement is to prevent runaway, civilization-destroying global warming—in other words, to create the necessary political and economic conditions for a last-ditch, all-out effort to keep enough fossil fuels in the ground—then isn't that work *already* about racial, economic, social, and generational justice? Because the consequences, I said, if we *don't* do everything possible to keep fossil fuels in the ground—

"Then we're not going to have *any* justice," Bullard interjected.

"In terms of the moral imperative," he added, "looking at the severe impacts—the impact on food security, on cross-border conflicts, war, climate refugees—when you look at the human-rights piece, in terms of threats to humanity, if we drew it out and looked at it, I think more people would be appalled at these little baby steps that we're taking. This is an emergency, and it calls for emergency action—not baby steps, but emergency action."

Nevertheless, Bullard also explained why that all-consuming focus on greenhouse emissions is insufficient by itself—and is at the heart of the tension between environmental justice and the climate movement.

"You have to understand that in order to have a movement, people have to identify with—and *own*—the movement," he said. "Just saying climate change is a big problem is not enough to get people to say, 'We're gonna work to try to keep coal and oil in the ground.' There has to be something to trigger people to say, 'This is my own movement.'"

Bullard believes that the climate-justice framework can "bring more people to the table." Take the example of coal plants, he said. "Moving away from coal, in terms of CO_2 and greenhouse gases—the environmental-justice analysis is that it's not just the greenhouse gases we're talking about; in terms of health, it's also these nasty copollutants that are doing damage right now. Not the future—right now." So to bring those people to the table, he continued, "you have to say: How do you build a movement around that and reach people where they are?"

~

In 2013, Bullard and his colleagues at TSU and other historically black colleges and universities—including Beverly Wright at Dillard and the Deep South Center in New Orleans—launched an initiative they call the Climate Education Community University Partnership (CECUP). "We're linking our schools with these vulnerable communities," Bullard told me, "trying to get to a population that has historically been left out. We're going to try to get our people involved."

Bullard noted that these colleges and universities have always had a special mission—Atlanta University (now Clark Atlanta), where Bullard began his graduate work in the sociology department created by W. E. B. Du Bois, was founded by the Freedmen's Bureau in 1865 to educate former slaves. Likewise today, he argues, "we should not run away from anything to do with justice and equity and opportunity." When you look at the most vulnerable communities, the "adaptation hot spots," he added, these are the same communities the schools were founded to serve, and often the very places in which they are located. "We're not going to wait for somebody to ride in on a white horse and say, 'We're going to save these communities!'" Bullard said. "We have to take leadership."

The initiative invests in a new generation of young scholars and leaders who can work at the intersection of greenhouse emissions, climate adaptation, and the classic environmental-justice issues of pollution, health, and racial and class disparities. "Our folks on the ground can make the connections between these dirty diesel buses, that dirty coal plant, and their kids having to go to the emergency room because of an asthma attack, with no health insurance," Bullard said. "We see it as human-rights issues, environmental issues, health issues, issues of differential power."

As I listened to Bullard, it was clear that anyone like me—with my privileged, big-picture view of the climate catastrophe—would do well to try seeing the concept of climate justice from the ground up, at street level, and through a racial-equity lens. Sitting down with five of Bullard's graduate students at TSU—and joined by two of his colleagues, sociologist and associate dean Glenn Johnson and environmental toxicologist Denae King—I was treated to a generous portion of that ground-up perspective.

For Steven Washington, a twenty-nine-year-old native of Houston's Third Ward and a second-year master's student in urban planning and public policy, "climate change means asthma; it means health disparities." Working in Pleasantville, a fence-line community along the Port of Houston, he was concerned about the city's notorious air quality, graded F by the American Lung Association, and what it means for a population—especially the elderly—ill-equipped to deal with impacts of climate change such as heat waves. For Jenise Young, a thirty-three-year-old doctoral student in urban planning and environmental policy whose nine-year-old son suffers from severe asthma, climate change is also about "food deserts" like the one surrounding the TSU campus—a social inequity that climate change, as it increases food insecurity, only deepens. (The wealthier University of Houston campus next door inhabits something of an oasis in that desert.) Jamila Gomez, twenty-six, a second-year master's student in urban planning and environmental policy, pointed to transportation inequities—the fact that students can't get to internships in the city, that the elderly can't get to grocery stores and doctors' offices, that the bus service takes too long and Third Ward bus stops lack shade on Houston's sweltering summer days.

I asked the TSU grad students if they saw the growing US climate-justice movement—especially the many college students and young people who want to foreground these issues—as a hopeful sign.

"My major concern is that this is a lifelong commitment," Young replied. "That's my issue with a lot of the climate-justice movement—that it's the hot topic right now. Prior to that, it was Occupy Wall Street. Prior to *that*, it was the Obama campaign. But what happens when this is not a fad for you anymore? Because this is not a fad."

Glenn Johnson, the coeditor of several books, including *Environmental Health and Racial Equity in the United States* (2011), chimed in: "It's a life-and-death situation. There are others who come into the movement, they have a choice—they can go back to their respective communities. But for us, there's no backing out of talking about the [Houston] ship channel. We are the front line; it's 24/7. When we wake up, we smell that shit."

"It's not one problem," said Denae King. "It's multiple problems—poverty, food security, greenhouse emissions, all of these things happening at once. In the mind of a person living in a fence-line community, you have to address all of the problems." Climate change is urgent, she added, "but still, I have to pay my bills *today*. I have to provide healthy food *today*."

All of which is undeniably true. And it is equally true that the overwhelming scientific evidence indicates that the window in which to take meaningful action on climate change is closing fast. Unless we—the United States and the world—act now, *today*, to begin radically reducing greenhouse emissions and building resilience, our children and future generations face impacts that will dwarf even today's worst environmental and economic injustices.

What you hear from climate-justice advocates working on the front lines—who understand this urgency perfectly well—is that precisely because of the emergency in which we find ourselves, the way to build the kind of powerful movement we need is to approach climate change as an *intersectional* issue.

After I left Houston that April, I spoke with Jacqueline Patterson, director of the Environmental and Climate Justice Program at the NAACP. One of the first things she did upon arriving in 2009, Patterson told me, was to write a memo looking at climate justice and the NAACP's traditional agenda. "It went area by area—health, education, civic engagement, criminal justice, economic development—and showed how environmental and climate justice directly intersect in myriad ways."

In the communities where she organizes, Patterson told me, "we see the links. The same facilities that are driving climate change are also causing immediate health and economic impacts in these very communities. So they have an added advantage to see coal be put out of business. They're the ones who have children stay home from school because of an asthma attack—or they're burying their children because of an asthma attack that wasn't caught in time. People are having lung disease who never smoked a day in their lives. And we talk about all of that in an intertwined way."

Patterson grew up on Chicago's South Side and graduated from Boston University before earning degrees in public health and social work from Johns Hopkins and the University of Maryland. She feels a spiritual pull to climate-justice organizing and is actively engaged with churches.

But her entry point to the work was her interest in how women are disproportionately affected by climate and environmental dangers: the spikes in sexual and domestic violence against women during and after disasters; the economic effects of women being primary caregivers for the sick and injured; the differential impacts of toxic exposures on women, from breast cancer to complicated pregnancies and birth defects. In 2007, she cofounded Women of Color United, and in 2009 partnered with the NAACP on the Women of Color for Climate Justice Road Tour. "Globally, it's very much a part of the conversation," she told me. "But here there's an absence of gendered analysis around climate change."

At the NAACP, Patterson's work rests on the understanding that if we're going to address climate seriously, then we're in for a rapid energy transition—one that's by no means guaranteed to be smooth or economically and socially just. In December 2013, the NAACP initiative released its "Just Energy Policies" report, looking state by state at the measures that can help bring about a *just transition* to clean energy. "In talking about such a major shift in such a major part of our economy, we're being very explicit that we're not just talking about renewable portfolio standards and energy efficiency standards and net metering policies," she said. "We're saying that each state needs to have 'local hire' provisions, at the state and local levels, and provisions for disadvantaged business enterprises—minority and women-owned businesses. We have to be very intentional about an economic-justice transition along with the energy transition."

The day before I talked with her, Patterson said, she stood next to NAACP leaders at a press conference in Milwaukee, "and we were talking about starting a training and job-placement program for formerly incarcerated youth and youth-at-risk around solar installation and energy-efficiency retrofitting." An energy-efficiency bill was recently introduced in the Missouri Legislature, she noted. "Before, we might not have seen the NAACP getting behind that legislation, because the energy conversation wasn't seen as part of our civil rights agenda. Now, we're in with both feet."

———

As I listened to Jacqui Patterson, and to Bob Bullard and his colleagues in Houston, and to Hilton Kelley in Port Arthur, a question that kept running through my mind was simply this: Where is the left? Where has it been? Why is this not at the top of the progressive agenda, with a robust social movement merging environmental, economic, and racial justice under the banner of climate justice?

It's an odd thing, if you think about it. For a long time, in many precincts of the left, and especially across a broad spectrum of what could be called the economic left, humanity's accelerating trajectory toward the climate cliff has been little more popular as a topic than it is on the right. In fact, possibly less so. Plenty on the right love to talk about climate change, if only to deny its reality, downplay its urgency, and take shots at Al Gore. On the left—to say nothing of the ever coolheaded center—denial takes different forms.

It's unclear what explains this reticence about the existential threat facing humanity, beginning with the poorest, including in the United States. But a lot of people I know in the climate movement think that the left, and the economic left in particular—pretty much the entire spectrum from mainstream liberals to anarchist Occupiers—has not yet taken on board the real implications of our galloping climate catastrophe. Not really. Not the full, stark set of facts. It's as though the implications of climate science, when you really begin to grasp them—for example, that the depth and speed of the necessary emissions cuts are incompatible with economic growth as traditionally defined—are simply too radical. Even for radicals. (There are exceptions, of course, but peruse back issues of leading journals on the left and you find that climate, for the most part, gets a passing mention—and when it is discussed, it's too often siloed, safely contained under an "environmental" rubric, as if to check that box for certain funders. When the labor correspondent starts writing about climate justice, then we know we're getting somewhere.)

The truth is, anyone committed to the hard work of bringing deep structural change to our economic, social, and political systems—the kind of change that requires a long-term strategy of organizing and movement building—is now faced with scientific facts so immediate and so dire as to render a life's work seemingly futile. The question is how to escape that paralyzing sense of futility—and how to accelerate the sort of grassroots democratic mobilization that's desperately needed if we're to salvage any hope of a just society.

At the same time, as I kept hearing in Houston, mainstream climate advocates who want to broaden the climate movement have too often been tone-deaf, if not completely absent, on issues of economic justice and inequality. How, then, to reconcile these two tendencies—the economic left's avoidance of climate, and the climate movement's avoidance of economic and social justice? How to merge these fights with the kind of good faith and urgency required to build a real climate-justice movement?

I don't know anyone who has all the answers, but I do know some people who are asking the right kinds of questions and looking in the right kinds of places. Naomi Klein, in 2014's *This Changes Everything: Capitalism vs. the Climate*, and James Gustave "Gus" Speth, in his 2012 book, *America the Possible: Manifesto for a New Economy*, have made the case that climate change could serve to unite the left in a movement of movements with economic justice and human rights at the core. And as I've engaged deeply in the climate movement, I've come to know people who are dedicated to that very challenge, starting the necessary conversations and actually working to connect climate with economic- and social-justice organizing across the country. As it happens, quite a few of them came out of the Occupy uprising. Many are involved with networks such as the New Economy Coalition, where Speth is among the core advisors, and the Climate Justice Alliance. (You'll read more about both of these below.) And what they're pointing to, it appears to me, is a promising convergence of climate justice and grassroots economic democracy, rooted in local communities and networked nationally. They may, in fact, be showing us what a new kind of movement—I think of it as *climate democracy*—can look like.

Every bit as important, they're acting with the kind of urgency and commitment that our unfolding catastrophe demands. They know there can be no climate justice without economic justice, but they also know there won't be any economic justice—or any justice at all, as Bob Bullard said to me—without facing up to our climate reality, simultaneously slashing emissions, making the fastest possible transition to clean energy, and working to build resilience in the communities where it's needed most. They know that the *climate* part of "climate justice" cannot be an afterthought, some optional add-on to please "environmentalists." Because the game is far from over, and no matter what happens in terms of national climate policy in the next few years—and the prospects are not pretty—current and future generations have to live through what's coming.

It was a weekend in late October 2013, and my friend Rachel Plattus was speaking to a roomful of college students and recent grads at the David L. Lawrence Convention Center in Pittsburgh, where they'd gathered along with some eight thousand other young activists at Power Shift, the biannual national convergence of the student climate movement. Rachel, who was then twenty-six years old, was the director of youth and student organizing for the New Economy Coalition, based in Cambridge, Massachusetts—a national network of more than a hundred organizations, large and small, with a shared commitment to (in the words of its mission statement) "a just transition to a new economy that enables both thriving communities and ecological health."

By Rachel's side was her good friend Farhad Ebrahimi, thirty-five years old, who served on the NEC board and who founded the Boston-based Chorus Foundation, which supports grassroots climate and environmental-justice organizing in communities around the country. I've come to be friends with Rachel and Farhad through the Boston-area climate movement, and I was tagging along there at Power Shift with them and their NEC colleagues. At first it was strange, I had to admit, to see Rachel and Farhad in front of a room at a high-tech convention center—during the previous year I'd been more apt to see them in church basements and community-organizing spaces leading nonviolent direct-action trainings, or on the streets engaged in protests against tar-sands pipelines and coal-fired power plants.

"I met Farhad at Occupy Boston," Rachel told the hundred or so young people who'd come to hear about the intersection of climate and economic justice (a strong showing, given the dozens of concurrent breakout sessions offered at Power Shift). "We spent a lot of time there a couple years ago, and it was a transformative experience for a lot of us."

Two important things came out of her Occupy experience, Rachel explained. First, she and several friends who had been "radicalized on climate issues," including Farhad, decided to form an organizing collective "to do resistance work around climate justice." At the same time, she began thinking seriously about the central question raised by Occupy but never really answered: "If you're so angry at this system, if all the people here have been wronged by the system, what are you proposing that we do instead?" While she and her friends wanted to keep organizing resistance, she said, "I found myself looking for a way to have an answer to 'What do you want instead?'" So, first, she dove into the worker-ownership movement in Boston and tried to start a worker co-op with some friends.

But also around this time, in late 2011 and early 2012, Rachel started talking with Bob Massie, a family friend, who had recently been hired to head the New Economics Institute (which merged with the New Economy Network in early 2013 to form NEC). A longtime social-justice and environmental activist (and, interestingly, an ordained Episcopal priest with a doctorate from Harvard Business School), Massie had served as executive director of Ceres, a large and influential network of environmental groups and institutional investors, and in 2003 he spearheaded the creation of its Investor Network on Climate Risk.

Once at NEC, Rachel began to realize that the kind of work going on in the "new economy" movement—with things like cooperatives and worker-owned businesses, community-development financial institutions, community land trusts, Transition Town initiatives, local agriculture and community-owned renewable energy, as well as efforts to reconceive corporations and redefine economic growth—is a way of challenging the dominant, extractive, and unsustainable model of corporate capitalism as we know it. Not simply rejecting that model or system but, as she emphasized to her Power Shift audience, "creating new economic institutions that are democratic and participatory, decentralized to appropriate scale so that decisions are made at the most local level that

makes sense and, rather than only prioritizing one thing—the maximiza-
tion of profit—prioritizing people, place, and planet."

"New-economy innovations are occurring all over the country, bub-
bling up," Massie told me. "What they lack is mutual awareness, mutual
support, and mutual connectivity." There's potential for real transformation, he believes, in providing those connections. "As people become
aware of each other, their frame of reference about what's happening, and
what could happen, changes. They realize all these problems are linked—
but all these solutions may also be linked." He pointed to what happened
earlier that year in Boulder, Colorado, where voters overwhelmingly ap-
proved a grassroots energy initiative to move the city from a big, corporate, coal-dominated utility, Xcel Energy, to a publicly owned municipal
utility that would expand renewables at the same or lower rates.

That's Boulder, of course, which is a long way socially and economically from Port Arthur and the Third Ward. Back in Cambridge, Rachel
and I spent a long morning at the NEC's start-up offices in Kendall
Square, near MIT, talking about these real-world challenges. I asked her
how she connects the new-economy work—which has genuine promise, at least in pockets around the country—with her work organizing
resistance to the fossil-fuel industry, often in solidarity with frontline
communities. First, she pointed out, "in a civil society that is essentially
owned by multinational corporations, driven to maximize profit over
all else, to engage in building these parallel economic institutions *is* to
engage in civil resistance."

But even more, she suggested, in the merging of climate justice and
economic democracy, it's the democracy part that may ultimately matter most. Rachel understands that the kind of deep, systemic change
envisioned by the new-economy movement (perhaps best articulated in
the work of Gar Alperovitz, the political economist and cofounder of
the Democracy Collaborative) is no doubt a long-term, evolutionary
process, on a timescale out of sync with our climate emergency. But she
argues that grassroots economic democracy, actually organizing to create
those alternative institutions, can also help build a base of political power
in the near term, at the local level, which is not only where all politics has

to start but all resilience as well—something we're going to need plenty of in the years ahead.

Rachel told me that she knows a lot of people who are focused primarily on the economic-democracy piece—and yet, she added, "almost all of them recognize the level at which that also plays into climate issues, how we build resilient communities." She pointed not only to something like the community-owned energy initiative in Boulder, but to projects like the Dudley Street Neighborhood Initiative in the Roxbury/North Dorchester area of Boston, which has brought a racially diverse, low-income community together around fair and affordable housing, community economic development, food justice, education, and youth empowerment. Also in Roxbury, Alternatives for Community and Environment (ACE), a well-established environmental-justice group with a strong emphasis on youth organizing and leadership development, has helped build coalitions across the city and region. Initiatives like these, Rachel said, are "building relationships, making sure the community is there, people interacting with each other in the kinds of ways we need people to be interacting with each other."

"Occupy did that, too," she said. "Being part of participatory democracy, in all its forms, does that: it gives people the skills and capacities they need" to help build a social movement.

That all sounded right to me. Indeed, Rachel was pointing to age-old verities of social movement building. And yet, I asked, where's the *climate* in that picture? What happens to communities like Roxbury and Dorchester—and West Port Arthur and Houston's Third Ward—where people are already struggling, if we don't urgently build the kind of grassroots power we need to shift the politics of climate, explicitly, and deal head-on with the emergency?

Rachel nodded. She knew exactly what I was getting at. After all, she had devoted many hours during the previous year to organizing against the Keystone XL pipeline—because of its importance to the climate fight.

"We have to be willing to tell the truth about what the dangers of climate change are, and how we balance immediate economic survival with longer-term survival," Rachel said. "We have to be willing to be honest about those things. But we also have to recognize when we're building power toward addressing the climate crisis, even if people aren't calling it the 'climate justice' movement."

In Pittsburgh, Farhad stood in front of the room wearing a gray hoodie with the words "Kentuckians for the Commonwealth" printed across it. He was talking about what he'd learned since diving into climate work in 2006 and seeing even the most inadequate climate legislation die in Congress in 2010—the last time any such national legislation has had a chance of passing. Like so many others in the climate movement at the time, he began to realize what was missing: "any sense of building political power, any sense of a social movement, and the intersectionality of climate justice and other social-justice movements." Through his young foundation, Chorus, he decided to start supporting grassroots organizing in frontline communities, those already bearing the brunt of the fossil-fuel industry—and one of the first places he went was Kentucky.

"We went to look at the extraction stuff going on, mountaintop removal," Farhad said, "and we saw that the folks who were trying to fight the coal companies, stop them from blowing up their mountains, were also doing great work around energy efficiency and renewables—and when it was tied together with this resistance work, it was actually much more effective."

He learned about Kentuckians for the Commonwealth, a statewide independent grassroots group that's been working for more than thirty years on democratic reform and economic and environmental justice (one Wendell Berry of Henry County, Kentucky, is among its strong supporters). KFTC does far more than work on coal and environmental health issues, central as those are in eastern Kentucky, where the group has its strongest base. Confronting climate change is the first plank of the KFTC platform, but much of its work is on local and regional economic development, tax-justice issues, mass incarceration and voting rights, as well as worker cooperatives, local agriculture, and community-owned and distributed renewables.

The folks at KFTC frame all of these as essential parts of a "just transition" from the old, extractive, exploitative economy to a new, more democratic clean-energy economy. The idea is that even as they build grassroots political power, they're also creating real economic alternatives to fill the void left by the coal industry. As executive director Burt Lauderdale explained to me, KFTC has established its presence in state politics. In 2010, as part of its strategy to move rural electric cooperatives away from overdependence on coal, the group helped prevent the East Kentucky Power Cooperative from building a new coal-fired plant and reached an agreement with the utility to explore energy efficiency and clean-energy alternatives. In 2013, KFTC convened the Appalachia's Bright Future conference, helping shape the agenda of a highly touted Eastern Kentucky "summit" that December, called by Governor Steve Beshear, a Democrat, and Republican Congressman Hal Rogers, to jump-start an economic transition in a region reeling from the loss of coal-industry jobs.

Kentuckians for the Commonwealth is part of the emerging Climate Justice Alliance, a national collaborative network of more than thirty-five grassroots and supporting organizations committed to uniting hard-hit communities on the front lines of both climate disaster and fossil-fuel extraction and pollution—Indigenous, African American, Latino, Asian Pacific Islander, and poor and working-class white communities. In 2013 it launched the Our Power Campaign, focusing on three "hot spots": in the Black Mesa region of the Navajo Nation, led by the Black Mesa Water Coalition; in Detroit, led by the East Michigan Environmental Action Council; and in Richmond, California, led by the Asian Pacific Environmental Network and Communities for a Better Environment. Each of these groups is working to address the local impacts of fossil-fuel extraction and infrastructure—coal mines and power plants in Black Mesa, a coal plant and oil refinery in Detroit, and the massive Chevron refinery in Richmond. At the same time, and just as important, they're applying principles of economic democracy to work toward more sustainable and resilient local economies in struggling communities.

Another member of the Climate Justice Alliance is the NAACP's Environmental and Climate Justice Program, headed by Jacqui Patterson. As she made clear to me, the idea of a just transition is "integral" to their work. "We don't talk about closing any coal plant without making sure that every worker has a different way to make their living. We consider tax revenues, the effect on the tax base of the communities where the plants are located, as well as other revenue streams impacted by that kind of transition."

Jihan Gearon, executive director of the Black Mesa Water Coalition (BMWC), grew up on the impoverished Navajo reservation in northern Arizona, where unemployment runs around 54 percent and approximately eighteen thousand homes lack electricity, despite the utility lines running over their heads. Black Mesa, sacred to the Navajo people, has two coal mines operated by multinational Peabody Energy, and is surrounded by five coal-fired power plants, creating air pollution that rivals big cities such as Denver while contaminating and depleting precious water. Gearon told me that their approach to climate is "holistic," addressing not only emissions as they've fought to shut down coal plants and mines, and hold Peabody accountable for environmental damage, but also adaptation—especially as water becomes scarcer—and sustainable economic development. "We are not content with parts per million of CO_2 reduced," she said. "We also want to ensure that we protect health, water, and jobs as we reduce CO_2." BMWC's major initiative— the industrial-scale Black Mesa Solar Project, envisioning a series of 20MW to 200MW solar photovoltaic installations—could be a model for how community sovereignty can be protected and enhanced through control of locally generated renewable energy. At the same time, BMWC is engaged in movement building beyond Black Mesa, engaging with the Climate Justice Alliance and establishing the Southwest Indigenous Leadership Institute in 2010, working with youth to develop a new generation of Indigenous leaders.

In the face of our climate reality, Farhad told me back in Boston, "economic transition is inevitable." In Appalachia and Black Mesa, as coal goes away, it's already happening. The question is, he said, "Will the transition be *just* or not?"

That question, I suggested, begged another one, equally important: Will the transition be *fast* enough?

Look, Farhad said, in any likely scenario, "what are we going to need, no matter what? Local political power and local resilience." We won't get where we need to be politically on climate change, at the national and international levels, "without real local base building." And if we *don't* get anywhere at the national and international levels—"well, then, we're going to need the local work in place so that we can take care of each other as the old way of doing things slips away."

In connecting resistance to resilience, Farhad said, "We're trying to go from 'no' to 'yes.' But it's gonna be a really fuckin' rough ride. It's gonna be a rough ride because of climate change, but it's also gonna be a rough ride politically and economically." Kentuckians for the Commonwealth, Farhad pointed out, is important right now because of how it intervenes in Kentucky politics, organizes communities, and fights the big coal companies. "And when the climate changes and what grows there changes and how they can live there changes—they're going to need that ability to act collectively to deal with all of that as well."

When Rachel and I talked that day in Cambridge, I asked if she agreed that much of the economic left had yet to take on board the full magnitude and urgency of the climate catastrophe.

"I mean, the *climate* movement has barely taken it on board," she replied. And yet, she said, how can she blame anyone? "Have *any* of us really taken it on board? It's like, how do you walk around with a knife in your chest? How do we begin to deal with the despair in such a way that it becomes useful?"

"There are a lot of folks, even in the climate movement," she said, "and certainly on the economic left, who haven't made the decision to take on the reality of it—and to recognize that this fight, which for them was never really about survival, all of a sudden is."

"It's interesting," Rachel said, "because there certainly are parts of the left, not the liberal elite, but parts of the left, for whom being engaged has always been about survival"—like those, she pointed out, who have fought their whole lives for racial justice. One of her heroes, she said, is the great black-freedom leader Ella Baker, who was instrumental in the creation of the Student Nonviolent Coordinating Committee, SNCC, in 1960. But before that, Rachel points out, "she came up in the movement building parallel economic institutions, building co-ops. And then went on to start the Freedom Schools, a parallel institution, and the Freedom Democratic Party."

"There is a deep, rich tradition of organizing for survival," Rachel said. "In fact, it's the only thing that's ever worked."

———

Beverly Wright knows about survival. A native of New Orleans, where she has deep roots—"I can trace my heritage back eight generations in this place," she told me—her experience of Hurricane Katrina was personally devastating. "That terrible storm," she writes in *The Wrong Complexion for Protection*, "washed away everything that I owned, everything that I had inherited, and every tangible bit of memorabilia that captured my life and family experiences." What African Americans experienced in the aftermath, she writes, "was a disaster that overshadowed the deadly storm itself." A decade later, the lack of policies to protect the city's most vulnerable threatens "a permanent and systematic depopulation and displacement of the African American communities of New Orleans."

The sordid history of Katrina—the government failure, the neglect and abandonment of the city's black population—is well known and documented, including by Wright and Bullard in *The Wrong Complexion for Protection*. But long before that storm and its aftermath, Wright herself had helped document the deep structural and environmental racism afflicting African American communities along the Mississippi River in Louisiana—and had drawn the connections to climate and climate justice.

"I was there," Wright told me, recalling the first climate-justice summit in The Hague, in 2000. "For me, defining climate justice and coming to terms with that whole concept was transformative. It evolved, for me, out of the environmental-justice movement—it was a kind of eureka moment when I realized these things were coming together."

For Wright, that process began at home. New Orleans, she points out, is in what's known as Cancer Alley. "But growing up," she told me,

"I had no idea that I was living in Cancer Alley." It was only after returning from graduate school at SUNY Buffalo, where she received her PhD in sociology in 1977, and teaching at the University of New Orleans, that she began to realize the extent of the industrial poisoning of her environment. "I began to hear about the chemical corridor," she said, "and that our cancer rate was extremely high, and people suspected it was related to the plants along the Mississippi River—which added up to about 136 petrochemical plants and six refineries on an 85-mile stretch of land." The number grows to 145 plants, she said, when you include the Calcasieu Parish corridor running west to Lake Charles, Louisiana.

After working with Bob Bullard on *Dumping in Dixie*, Wright was asked to testify at a 1992 civil rights hearing in Baton Rouge involving affected communities along the Mississippi, and in preparation she decided to investigate personally. "So I got in my little van, and brought two students with me, and drove up and down the Mississippi River corridor," she said. "What I saw was just appalling. I saw communities that were fence-line to these facilities. I got headaches." Most disturbing, she told me, was the vacant land. "I would ask people, 'What was this over here?' And they'd say, 'Well, that's where the white people lived. The boss told us he was going to come back and buy us out, too, but he never did.'"

"In the South," Wright explained, "black and white people, though segregated, often live very close to one another. But we would see no white people—just these facilities and black people, because the white people had been bought out."

Wright took her observations to the EPA, she told me, and in 1994 the Deep South Center (which she had founded in 1992) received a grant to study the spatial distribution of facilities by income and race. It became an eight-year project. "The data was not easy to get," she said. "You could get the income data but you couldn't get the race data." Ultimately, they produced the first GIS map of the Mississippi River chemical corridor. "We were able to show that African Americans live closer to these chemical plants—that eighty percent of them live within three miles of a TRI [Toxic Release Inventory] facility. We looked at the data twenty years later and it had gotten worse."

In the 1990s, as Wright, Bullard, and colleagues began to look at the global impact of the fossil-fuel industry on communities of color, they saw the connection to climate change—the same industry that was destroying Indigenous peoples' lands and poisoning black and brown people around the world was not only driving global warming but was blocking efforts to deal with it at the national and international level. At the 2000 summit in The Hague, they talked with representatives from Africa and Asia and island nations. "We began to see what communities would be most affected by climate change," she said. "And even before Katrina, we began to see that because of where we live, we look a lot like these island states, in terms of sea-level rise." And not only sea level, but the concentration of petrochemical infrastructure, and its vulnerability to intensifying storms, means that the frontline communities face a kind of "double jeopardy," as Wright calls it. "The same petrochemical plants that are poisoning them are causing things that could wipe them off the face of the earth."

When I asked Wright how Katrina helped shape her understanding of climate justice, she said it came down to two things: "Who is prepared, and then who actually recovers." In their chapter on the Katrina recovery, she and Bullard write of the "second disaster" often experienced by the poor and people of color: "Prestorm vulnerabilities limited the participation of thousands of low-income communities of color along the Gulf Coast in the poststorm reconstruction, rebuilding, and recovery. In these communities, days of hurt and loss are likely to become years of grief, dislocation, and displacement." In a tone of biting irony, Wright and Bullard go on to offer an all-too-real "Twenty-Point Plan to Destroy Black New Orleans"—from "Selectively hand out FEMA grants" and "Redline black insurance policyholders" to "Promote a smaller, more upscale, and 'whiter' New Orleans," "Delay rebuilding and construction of New Orleans schools," and perhaps most damning of all, "Hold elections without appropriate Voting Rights Act safeguards." On the latter, they note that in November 2005, three months after Katrina, 80 percent of New Orleans voters were still displaced—and that African Americans made up two-thirds of the city's population before the storm. Nearly three-quarters of polling places had been damaged or destroyed. Holding elections under these circumstances, they write, "is unprecedented in the history of the United States, but also raises racial justice and human rights questions."

Wright still saw the antidemocratic effects of the Katrina disaster nearly a decade later. So many people had been displaced within the city, forced to move from low-income housing projects that were never rebuilt, that a crucial sense of community was being lost. "It's like they're living in a foreign country," she said. "You've ripped up these families who've been in these neighborhoods, pushed them out to the east, which is another world. You don't have the community cohesion, a feeling of belonging, and you've seen the voting levels go down. They no longer have Miss Johnson who would walk the community and say, 'Today is voting time! Get out and vote!'"

I asked Wright if she sees that sense of community, now threatened, as an essential part of what makes for resilience. I mentioned that we often hear how resilient the people of New Orleans have been.

"This bothers me a lot," she said. When people talk about the "resilience" of the African American community in New Orleans, she told me, "The question really is, 'How did you survive this? What did you do? My God, that was just fabulous that you were able to survive that.'" But in the whiter, more affluent parts of the city, she said, "resilience has to do with how they're preparing people for another event. What should we be telling people to do with their houses? How can we fix these streets so there's less flooding? It's about building an infrastructure so that a community can be resilient. But when they talk about us, it's always, 'How did you survive?' It's two different stories."

A decade after Katrina, Wright told me, advocates and organizers like herself still didn't have the kind of institutional support from funders that they needed. "We need researchers, a research policy center, so that we can go before the city council and demand the proper infrastructure. We need organizers on the ground, so we can organize these people once again to vote." Nevertheless, she insisted that the situation could still be an opportunity for the African American community in New Orleans— "if we organize the way we should."

~

In his spacious, sunlit office at Texas Southern University, with its view out over the modern campus buildings and the streets of the Third Ward, Bob Bullard talked about growing up in the small and deeply segregated town of Elba, Alabama. His parents—his mother a housewife, his father a laborer—believed in education, and he graduated from high school there in 1964, the year of Freedom Summer and the Civil Rights Act, and went on to Alabama A&M, the historically black university in Huntsville. He graduated in 1968, and served in the marines for two years, but was spared from going to Vietnam. "I was lucky."

Bullard was formed by the civil rights struggle. "I was a sophomore in 1965," he said. "That was the year of Selma and the bridge. As students, you're very conscious." He told me he revered not only Martin Luther King Jr. but Ella Baker, Rosa Parks, Fannie Lou Hamer, Malcolm X, Stokely Carmichael, John Lewis. "There were a lot of them—and these people were very young. You identified with a struggle, and you saw it as *your struggle*."

Bullard has written about Dr. King's final campaign, in 1968, when King went to Memphis to march in solidarity with striking sanitation workers. It was a battle, he notes, where everything came together. "It was about worker rights," he told me, "it was about civil rights, it was about health equity—because garbage workers were working in conditions that were totally inhumane." And it was about environmental justice. "I tell my students, if you don't think garbage is an environmental-justice issue, you let the garbage workers go on strike."

"All these things intersected," he said.

And today, with our pressing issues of inequality, he said, King's emphasis in the end on poverty and economic justice appears prescient—and all the more relevant. "Who knows, if King had lived," Bullard mused, "with the Poor People's Campaign—and now we talk about growing inequality in our nation."

If environmental justice emerged out of the civil rights struggle, then you could almost say that Bullard's work, and the movement to which he's dedicated his life, began there in Memphis—picking up where King's work was cut short. "Clearly, the Memphis struggle was much more than a garbage strike," Bullard writes in the autobiographical essay that opens *The Wrong Complexion for Protection.* "The 'I AM A MAN' signs that black workers carried reflected the larger struggle for human dignity and human rights."

———

Hilton Kelley drove me up Houston Avenue, through what he calls Old Port Arthur, parallel to the railroad tracks that separate the African American west side from downtown. "This was the booming area during the heyday of Port Arthur," he told me. He pointed to a vacant lot on the near corner. "That was called Antoine's Auditorium, it was on the Chitlin's Circuit. Ray Charles performed there, Al Green, Aretha Franklin—all the greats. They knocked the building down about seven years ago." As we drove alongside the tracks, Kelley pointed to at least three small grocery stores that had long since gone out of business.

It's not the neighborhood Kelley remembers. In *A Lethal Dose of Smoke and Mirrors*, a memoir he published in 2014, Kelley writes about the decision to move back to his hometown. On a return visit from California in 2000, he was not only disturbed by the unusually high rates of cancer and other illnesses in the community. "Other things began to emerge," he writes. "The vacant properties now stood out in the midst of dilapidated buildings, businesses that were once there were now gone, and playgrounds were almost non-existent." When he was a kid in the sixties, "we had grocery stores, restaurants, small hotels, and even a pharmacy." Growing up, he writes, "we had a YMCA, an Olympic size swimming pool . . . after school programs. But no more, all of that was gone."

When he got back to California, he couldn't rest, and he started making lists of things that people should be doing, but weren't, to revive his Port Arthur community. One day, he writes, he finally looked in the mirror and said to himself: *"You keep thinking and talking about what nobody is doing in Port Arthur, but what are you doing? You're from Port Arthur, why don't you do something?"*

We crossed the tracks and drove past a housing project built in the 1970s. A few blocks away, Kelley showed me Saint John's Missionary Baptist Church, where the Reverend Elijah "EJ" James had allowed him to hold some of his first organizing meetings. But he's been asked not to distribute fliers outside some of the churches, Kelley said. He affected an old man's voice: "We can appreciate what you're doing, son. But don't pass that out around here." He added, "Some of them work at the plants."

"There are still folks who won't speak up because they're tied to those industries by their pensions or because they work there, or their parents or their kids work there." But he explains to them: "Just keep in mind, the toxic air that you're breathing when you go out there to work, when you come home you're still breathing it. But not only are *you* breathing it, *your baby* is breathing it. Think about those little lungs, they're more susceptible to the poisons."

We stopped to see his old high school, now a middle school, and I noticed the flag was flying at half-staff and wondered why. We both thought for a moment.

"Oh, it must be for MLK," Kelley said.

Of course. I had completely forgotten—it was April 4.

"I remember when Martin Luther King was shot," he said. "You could hear the neighbors crying. So I ran down the street to tell my mother, who was down at the Laundromat, and she was already in tears. She'd already heard about it. I was seven years old. It was a sad day."

We drove down Fourteenth Street, past the small houses—some in good repair with well-kept front yards, many others in poor condition, some at the point of collapse. A few blocks farther, where the road ends, was Carver Terrace, the housing project where Kelley grew up, alongside the Valero refinery fence. Carver Terrace was empty now, slated for de-molition, its residents given housing vouchers with the option to relocate to a new project in another part of town—one at least not directly in harm's way. The last family had moved out about three weeks earlier, Kelley told me.

We got out and stood among the rows of long, plain-brick, two-story buildings. "If you'd come here six months ago," Kelley said, "you would've seen kids running across the street and playing ball right here."

I asked him how it felt to see it like this now.

"Oh, man, it's like *The Twilight Zone*," he said. "I'm getting used to it, but I ride by here every day, just remembering. My grandmother, my great-grandmother, used to live out here. My friends, a lot of my family, a lot of them stayed here—I came back twenty years later, and they were still out here."

"We used to stay here with my mother," Kelley said, pointing to one of the units. "When she first broke away from the projects, we were little kids, four or five years old. She always wanted to get away from here, she wanted to be independent, stand on her own. She had that pride. But we had to come back, and we moved into this place here. The one where I was born is down here." He showed me another long row of brick buildings. "Row 1202, the third one, 1202E. That's where I was born, right behind those brick walls."

"It's bittersweet," he said, "to see this place quiet and deserted like it is. But it's necessary, because it's time for people to get out of harm's way. And they're going to turn this area into a green belt."

Perhaps fifty yards from where we were standing, and even closer to a playground with colorful new play structures, exposed pipes emerged from the berm along the refinery fence. Signs read: Warning: Light Hydrocarbon Pipeline.

Kelley drove me by the new health clinic and the community center, both built as a result of CIDA's relentless pressure. He told me that he never thought he'd be doing this work for as long as he has. "But here I am," he said, "fourteen years down the road, still chopping away at it. New issues keep cropping up. But trust me, I'm no ways tired."

"What I've discovered," he said, "is that we are a necessary entity in this community. I'm here to stay."

He told me about a new performing arts and education center he wanted to open in a small building near Kelley's Kitchen that he'd recently purchased and was renovating. The property downtown is dirt cheap, he said, and he dreamed of creating a place where people from the community, especially young people, could gather—and bring some new activity to the downtown area.

Young people have always been a priority for Kelley and for CIDA, providing after school activities and tutoring, "keeping kids off the streets." Recently, he said, young women from a Port Arthur alumni group had invited him to speak to their organization. "They honored me with a little certificate, said how proud they were of the work I'm doing, and wanted to encourage me to keep going, because I'm one of the loudest advocate voices out there, and they appreciate what I'm doing. That really made me feel good, for young folks to show their appreciation that way, and to know they're paying attention. That's huge."

It was a beautiful day, and Kelley drove with the windows down. A middle-aged woman on the street called out to him.

"How's it going?" Kelley said, genuine warmth in his voice.

"Pretty good," she called back. "How you doin'?"

"I'm hangin' on in there, enjoyin' this day."

"This is a great community to grow up in," Kelley told me. "I ran and played up and down these streets. I love the smell in the air right now, the plants growing, the springtime. We've got a pretty good day today—don't have any high emissions levels. I'm lovin' it. You can smell the flowers."

———

The next morning, I went back on my own and drove around downtown and the west side of Port Arthur. It was overcast now. The gray light altered the mood of the day before, and I was overcome by a need to see the ocean, across Sabine Lake and the coastal marshes on the Louisiana side.

So I drove out of Port Arthur on Highway 82, passing still more petrochemical plants along the way, and stopped after half an hour at a row of beach houses built high on sturdy pilings. I stood on the strip of sand, and the Gulf of Mexico lapped at my feet. The wind on my face came fresh and welcome—but on the horizon, all up and down the coast, were the platforms. There was no escape. I got back in the car, turned the ignition. No escape.

Heading back into Port Arthur, crossing the wide channel at the mouth of the lake, I drove over the Martin Luther King Jr. Memorial Bridge. As I crested its steep ascent, Port Arthur came into view, the Valero and Motiva refineries spread out in front of me. The dystopian petrochemical landscape stretched into the distance—and I caught my breath at the sight of it as I descended.

PART TWO

FOUR

We Have to Shut It Down

If ever truth reaches power, if ever it speaks to the individual citizen, it will not be the argument that convinces. Rather it will be his own inner sense of integrity that impels him to say, "Here I stand. Regardless of relevance or consequence, I can do no other."

—AMERICAN FRIENDS SERVICE COMMITTEE, *Speak Truth to Power: A Quaker Search for an Alternative to Violence,* 1955

We mean to speak now with the weight of our whole lives.

—CREW OF THE *Golden Rule,* IN A LETTER TO PRESIDENT DWIGHT D. EISENHOWER, JANUARY 9, 1958, STATING THEIR INTENTION TO SAIL INTO THE PACIFIC NUCLEAR TESTING ZONE

A stone's throw from the entrance to the Brayton Point Power Station in Somerset, Massachusetts, the pavement forks down to the left and dead-ends at a forlorn strip of public beach alongside a brave remnant of wetland, beyond which are small houses. Directly across a narrow inlet from the beach, the power plant rises, half a century old, with its towering smokestacks and the long pier where thousands of tons of coal are regularly unloaded from giant freighters. Not only do the people who live in the neighboring community breathe the pollution from all that coal, but Brayton Point is the largest coal-fired power plant in New England and has long been among the largest sources of carbon emissions in the Northeast.

On a bright, calm morning in the early spring of 2013, I stood on that beach with Ken Ward, a fellow climate activist I'd met through 350 Massachusetts. Ken, the father of a thirteen-year-old boy, had turned fifty-six the previous November and was then living in Boston's Jamaica Plain, where he supported himself as a carpenter and handyman. He's a good-humored, naturally buoyant kind of guy, with a mop of dark hair and a salt-and-pepper mustache. He's studied at Andover Newton Theological School and likes to play Dylan songs on the mandolin. He's also a veteran environmental insider—cofounder of the National Environmental Law Center and a former deputy director of Greenpeace USA—who happens to be a sharp critic of mainstream environmentalism, which he argues has failed to grapple seriously and urgently enough with the clear and present threat of catastrophic climate change.

Ken brought me down to that beach so that I could see the view. On the water in front of us, the hulking, black-hulled freighter *Energy Enterprise* rested at the pier, carrying some forty thousand tons of West Virginia's finest. An infernal mountain of the stuff rose behind it. Ken wanted me to stand there so I could picture for myself the sheer physical mass of the power plant, the ship, and the coal—and so I could see where, in some two months' time, Ken and a thirty-one-year-old Quaker activist named Jay O'Hara, from nearby Cape Cod, would put themselves in the way of that ship.

Early on the perfectly clear morning of May 15, after an impromptu sunrise prayer service with two members of their support team on the docks in Newport, Ken and Jay motored north to Brayton Point in a 32-foot wooden lobster boat—which they'd acquired and rechristened the *Henry David T.*—flying an American flag and a banner that read #CoalIsStupid. They were about two hours ahead of the *Energy Enterprise,* and Jay, skippering, positioned the lobster boat in the ship channel along the pier—right where the 689-foot freighter would have to dock and unload. Intending to stay a while, they proceeded to drop a well-fastened, 200-pound mushroom anchor off the stern of the *Henry David T.* Ken called the Somerset police and said they were there to carry out a peaceful protest.

Sometime before 11:00 A.M., the *Energy Enterprise* came into view, followed close behind by multiple high-speed Coast Guard boats. As the freighter bore down on Ken and Jay, the ship's captain made radio contact, ascertained their intentions, and advised them and the Coast Guard that he had ordered "defensive measures" on deck and was prepared to "protect" his crew. Meanwhile, from somewhere above them on the pier, Ken and Jay heard the distinctive *chck-chck* of a rifle, chambered and ready. When the freighter finally came to a stop, its prow practically loomed over the lobster boat. Coast Guard personnel boarded the *Henry David T.*, conducted a safety inspection, and calmly took control of the situation.

On the website they'd created for the protest, shared via social media around the world, Ken and Jay explained the reasons for their action. Those reasons boil down to this: even the most politically ambitious plans to address climate change at the national level, including President Obama's strategy of executive action to impose limits on power-plant emissions, fall far short of what the scientific consensus says is necessary if we're to have the slightest chance of averting global collapse. But if we accept what climate science is telling us—that humanity faces an existential threat and that we've all but run out of time to address it—then we should start acting like it. This means, first and foremost, that we have to stop burning coal, whatever the cost—because the cost of continuing to burn it is immeasurably greater. In a manifesto posted on the website, Ken and Jay wrote:

> To lose the world on our watch is a miserable prospect. To lose the world when a solution is available is perverse. Denying outright that climate change exists is the most extreme response, but considering climate change to be anything other than the single most important matter facing humanity has the same effect.
>
> What we need to do is relatively simple. Whether there is time to avoid the tipping point, we don't know, but that shouldn't prevent us from making the best possible effort.
>
> First thing: stop burning coal.

They concluded:

> We are faced with an imperative like none confronted by any
> previous generation. . . . It is our choice to take direct, nonviolent
> action—putting our bodies between the Brayton Point coal plant
> and its water-borne coal supply—in an attempt to achieve the
> outcome necessary for planetary survival: the immediate closure
> of Brayton Point Power Station.

By the time that two-hundred-pound anchor was hauled up from
the channel bottom by a salvage crane, as the sun went down behind the
plant, Ken and Jay had managed to block, for a day, the delivery of those
forty thousand tons of coal.

———

When I told Ken that the action at Brayton Point reminded me of something Greenpeace might have done, back in the day, he just smiled:
"Yep."

In fact, if that image of a lobster boat blocking a coal freighter called
anything to mind, for me it was the classic Greenpeace footage, shot
in the Pacific waters off the coast of California in June 1975, in which a
small band of committed souls in Zodiac inflatables positioned themselves between Russian whaling ships and a fleeing pod of sperm whales.
Those iconic images—small boats against a massive industrial force, courageous individuals putting their bodies on the line in a quasi-mythic
David-versus-Goliath tableau—exploded into popular consciousness
like a "mind bomb," to use the favorite phrase of counterculture journalist and Greenpeace cofounder Bob Hunter, who was one of those in
the Zodiacs.

And yet those grainy frames of longhaired environmentalists saving
whales can be somewhat misleading. When you look into the deeper
origins of Greenpeace, as historian Frank Zelko notes in *Make It a Green
Peace! The Rise of Counterculture Environmentalism* (2013), the combina-

tion of "green" and "peace" in the group's name was meant to carry real significance. Nevertheless, Zelko observes, just as we put the emphasis on the first syllable of the name, so the "peace" half tended to be overshadowed as Greenpeace went on to become the world's most recognizable *environmental* organization.

But "peace" wasn't always the junior partner. Quite the opposite. Some of those most instrumental in the formation of Greenpeace in the early 1970s in Vancouver, British Columbia, had been deeply involved in the American Quaker peace and antinuclear movements of the fifties and early sixties. Indeed, the Quaker and Quaker-inspired pacifist voyages of the sailboats *Golden Rule* and *Phoenix*, protesting US nuclear tests near the Marshall Islands in 1958, were forerunners of the early Greenpeace boats, which ventured into the Pacific nuclear testing zones between 1971 and 1974. Greenpeace cofounders Irving and Dorothy Stowe and Jim and Marie Bohlen, heavily influenced by the Quaker practice of "bearing witness" and the emphasis on nonviolent resistance, inspired by Gandhian satyagraha, left a lasting imprint on the group.

Despite the media infatuation with the radical hippie ecology of Hunter and others, for these Quaker-influenced cofounders, saving humanity from itself was just as important as saving whales, seals, or any other species. And whatever tendency there may have been toward a holistic mysticism among the counterculture greens, those elder peace-movement veterans viewed their antinuclear protests, rooted in both science and moral conviction, as sober and eminently rational affairs.

Canadian journalist and Greenpeace cofounder Ben Metcalfe may have captured that founding spirit best. "We do not consider ourselves to be radicals," he wrote for a radio broadcast aboard the very first boat called *Greenpeace*, in September 1971, on a voyage launched by the group's immediate forerunner, the Don't Make a Wave Committee, to protest a planned US nuclear test on the Aleutian island of Amchitka. "We are conservatives," he declared, "who insist upon conserving the environment for our children and future generations of men. If there are radicals in this story it is the fanatical technocrats who believe they have the power to play with this world like an infinitely fascinating toy of their own."

"The world is our place," Metcalfe explained, "and we insist on our basic human right to occupy it without danger from any power group."

Like Metcalfe and comrades, Ken and Jay defy easy labels. However much their protest reminded me of a classic Greenpeace-style action, it would be a mistake to lump them in simplistically with the sort of "save-the-whales" environmentalism those iconic images may suggest. If anything, Ken and Jay are more like the Quaker-inspired, antinuke Greenpeace founders. They want to save creation from humanity, and humanity from itself.

———

The week before their action at Brayton Point—though at the time, I didn't know when, or exactly how, or even whether they would go through with it—I sat down separately with Ken and Jay for long, honest conversations. I wanted to know what it really was that drove them to take the kind of action they were prepared to take.

Ken doesn't think of himself as radical. "I think I'm a deeply conservative person," he told me, unconsciously echoing Ben Metcalfe aboard the original *Greenpeace*. "My politics are deeply conservative. I'm just a classic, democratic American liberal."

That's deeply conservative? I asked.

"Yes," he said, "I think it is. In the sense of the founding fathers. You can talk about radical tactics, I suppose, but if you're asking, 'What are you? Are you a radical?' Well, yes—I would like to radically break from what we're doing right now, but only to return to things that I think are classic American political verities and virtues. I'm not trying to create some utopia or some new thing."

We were having lunch on the terrace of a Panera café in Wayland, surrounded by a giant parking lot and the constant noise of construction, as stores and restaurants were being built for a new outdoor mall. Commerce being one of the classic American verities, I asked Ken what he wants if *not* some new thing.

"I think we have to figure out how to be a society," he said, "and how to be human, within a set of constraints imposed on us from the outside." For Ken, this means recovering what environmentalism was always supposed to be about. "One of the things we fucked up was allowing environmentalism to become Democrat, left, and partisan. Because it didn't use to be that way. Environmentalism, at its start, was beyond ordinary politics."

"If you ask what is the core of environmentalism," Ken went on, "well, I'll say respect for the community of the earth as a whole, rather than just species advocacy. It's an awareness that we're on a track to destroy everything, including ourselves, and that we need to solve these things—otherwise, there's no future. That requires you, if you accept that, to think, 'How do you solve those problems?' As a way of looking at things, it's fundamentally distinct from any leftist thinking."

One might say that Ken has a right to talk about environmentalism and left politics. Born in Marion, Illinois, and raised in Providence, Rhode Island—where his father, Harold R. Ward, a professor of chemistry at Brown, created the first environmental center at an Ivy League school, one of the first in the country—Ken spent his early career, beginning at Hampshire College in the mid-seventies, building the Ralph Nader-inspired Public Interest Research Groups (PIRGs) at the state and national levels, working to move them in an environmental direction. In the nineties, in addition to cofounding Green Corps (the first and only national field school for environmental organizers) and the Environment America network, he cofounded and was president of the National Environmental Law Center and served as deputy director of Greenpeace USA, running its day-to-day operations, from 1996 to 1998.

But the Greenpeace experience was "wrenching," Ken told me. "I thought that Greenpeace was central to how we could succeed, and yet they had shifted off of what was successful." A policy decision had been made, Ken said, that effectively stopped high-profile direct action campaigning. Meanwhile, he said, "we were barely working on climate."

It was in this context, in 1997, that Ken was "knocked completely off kilter," as he recalls, by a conversation with colleagues Jon Hinck, the campaigns director of Greenpeace International (who had helped build Greenpeace USA and served as its executive director), and Bill Hare, who was the climate policy director for the international organization.

"I was sitting in the Greenpeace International offices in Amsterdam," Ken told me, "and Jon Hinck is walking through the basic chart of the carbon dioxide going up. I knew all this stuff, but Greenpeace had just come out with a brilliant report, written by Bill Hare, called *The Carbon Logic*, which showed that to meet even the modest objectives of the IPCC's second report [1995–96], most known fossil-fuel reserves would have to stay in the ground." Hare wrote in the paper: "This report shows the implications for overall fossil fuel use, in the form of a 'carbon budget,' over the next century if the global community is to prevent dangerous climate change." It would only be possible, Hare continued, "to burn a small fraction of the total oil, coal and gas that has already been discovered, if such dangerous changes are to be avoided. Even the reserves of fossil fuels that are considered economic to recover now, with no advances in technology, are far greater than the total allowable 'carbon budget.'" That was written in 1997, more than a decade and a half before the IPCC would incorporate the concept of a global carbon budget into its *Fifth Assessment Report*.

Ken remembers his reaction to the idea that the vast majority of fossil-fuel reserves would need to stay in the ground. "I was like, 'What? Wait a minute. That's a very different sort of a problem here.' It struck me as an imperative and an opportunity for environmentalists. And so I went back to the US, where we had, I don't know, something like twenty-five campaigners working on toxics, and *one* on climate." He tried to change that, he told me, "but it became very clear that the things I thought needed to happen weren't going to happen."

Ken went back to the PIRGs, where he cofounded Environment America, but by this time he was exhausted, burned out. He took some time off and studied at Andover Newton Theological School. Raised a liberal Protestant in the Congregationalist tradition, Ken is a sincere Christian—and his faith, he says, is a practical one. Years before going to Greenpeace, he had decided to quit drinking, and the daily spiritual practice of prayer and meditation helped him stay sober.

While he was studying theology, his then-wife, a labor organizer with Service Employees International Union, became pregnant with their first and only child. When their son, Eli, was born in 2000, they flipped a coin to see who would be the at-home parent, and Ken won the toss.

It was around this time, at home with his baby boy, that Ken began reading the climate science in earnest. It seems the combination of becoming a father and delving into the science flipped a switch in Ken's brain: he could no longer view climate as merely one environmental issue among many. It was the only thing that really mattered.

Ken was struck, in particular, by the work of James Hansen, the longtime director of NASA's Goddard Institute for Space Studies, often called the country's top climate scientist, who famously testified before Congress in 1988 about the already clear threat of human-caused climate change.

"I remember distinctly," Ken told me, "I went to my dad and said, 'Who is this Hansen guy? Should I believe him?' Because you read the stuff and you go, we're really fucked."

"I went back around to all my friends, and I said, 'Hey, it's a lot worse than we even think. In fact, it's happening in our lifetime. We have to do this *now*.'"

In 2005, Ken set out to create a center, or network, through which to change environmentalism from within, to help reorient the big green groups toward a coordinated, all-out effort on climate and a comprehensive strategy commensurate with the scale and urgency of the threat. The kind of strategy that, Ken noted to me, still isn't on the table.

He called the initiative Bright Lines, and over the next few years he engaged in discussions with a large network of peers and old colleagues, including senior environmental organization leaders and strategists. The pitch for the Bright Lines network, as he framed it for me, was simple: "We were proposing a cheap, obvious thing to do. We've got this looming problem, it's beyond anything we've imagined, and really, what we're doing now isn't going to touch it. Let's at least have a place where we can talk about this stuff internally." But for whatever reasons—political, institutional, perhaps personal—the Bright Lines concept didn't take hold. "We had a network of two hundred to two hundred and fifty people. We had a series of meetings. But it was all based on having enough money to have a handful of people to do this stuff, and I couldn't raise a dime." Greenpeace was on board at one point, he said, but then backed out.

"People toyed with it, and they just didn't want to . . ." He considered his words. "There was no felt need for it."

"I mean, I wasn't entirely the best person to do it," Ken said, "so it was easy to cast it as 'me.' You know, this guy is an arrogant loose cannon, trying to dictate to us. Partly, that's right. If I had no arrogance I couldn't possibly have said those things to begin with."

Jon Hinck is now a lawyer and city councilor in Portland, Maine, where he has also served as a state legislator. He was among those in conversation with Ken during the Bright Lines period, and I asked him if he thought Ken's sense of where the environmental movement needed to go had been vindicated.

Yes, Hinck told me. "I agree with Ken—*entirely.* What he keeps saying, and I think it's fairly simple, is that you try to match your objectives and your strategy to the science that describes the problem." Even when the big environmental groups have focused on climate, Hinck said, pointing to the 2009–2010 cap-and-trade debacle, "the goals they're setting are too modest, with respect to the challenge faced. You have to ground it in the challenge faced, or it's meaningless."

"The magnitude of the climate problem is, unfortunately, too easily understated," Hinck said. "And I give Ken a lot of credit for just stepping out and articulating it."

Ken summed up his Greenpeace and Bright Lines experience succinctly. "Environmentalism, as we organized it," he told me, "has proved an inadequate tool for handling climate, because the scope of the problem, and the scale of the solution, can't be fit into the politics-as-usual framework within which our *organizations* prosper and careers are built. We forgot what we built the institutions for."

With Bright Lines struggling to get off the ground, Ken suffered a deep personal blow when he lost his closest friend and collaborator to alcohol and drugs. "She died, she was my former love," Ken told me. "When I was a child, when I was twenty, she was my first organizer."

It was around this time that Ken went through his most difficult period.

"I'll tell you the very worst moment in all of this for me," Ken said. "I was sitting in a psychiatrist's office in Hingham, having explained that I was having trouble functioning. This must have been 2005. And I'm sitting there, and this psychiatrist goes, 'So, let me see if I can get this story straight. You're saying that the world is ending, and that you have a view of what should be done to prevent this from happening.' And I say, 'Yeah.' And he goes, 'And nobody's listening to you, nobody's paying any attention to you.' And I go, 'Yeah, more or less.' And he goes, 'Is anybody aware of this? Is there anybody I would've heard of, who shares your view?' And I go, 'Yeah, James Hansen'—this is before he's doing anything political—'Hansen is the person who's identified the problem. He's a NASA scientist, you can read him. And Bill McKibben and Ross Gelbspan are the only other people I know who have a comprehensive view.' And he goes, 'So you're working with them?' And I go, 'No, I kind of disagree with them.'"

Ken smiled.

"And he goes, 'OK, I think what we're dealing with here is a classic case of bipolar mania, which spirals out of control, and if we don't immediately arrest this, you could be one of those people who's institutionalized, because we don't know what else to do with them, because they think they're Jesus Christ.'

"And I go, 'I don't really see how you get that.'

"And he says, 'Well, just listen to what you're saying to me. You're saying, effectively, you know a way to save the world, and nobody else does, and the only two people who might agree with you, you don't agree with them.'

"And I said, 'I'm not saying I know how to save the world. I have a view of what I think environmentalism should do. But then, I've been a senior leader. I mean, it would be like, if there's a banking crisis, and somebody had been a banker, and they looked around and said we should do the following, in monetary policy, would you call them—?'

"He said, 'I don't know what to tell you, but that's my firm opinion.' And I went to family and friends, and they said, 'Yeah, you have been acting pretty manic.' So I agreed to be put on a treatment of lithium, which makes you drool. And I remember sitting in a job interview, try-ing to not drool. And after a while I go, this is really fucked up. I'm sorry, this is not right. I do think I have a reasonable critique here. Just nobody's listening."

Ken is nothing if not resilient. In 2008, he turned to grassroots activ-ism, trying to help build the climate movement from the ground up, engaged with the burgeoning 350.org network in Massachusetts.

But Ken's worldview has darkened—or hardened. In the near term, "nothing is going to change," he told me. "We're too invested—there's too much power and money and lethargy. We're not going to do what we need to do, which is to shut all the coal plants down tomorrow."

"The best that we can hope for," he said, "if people are trying to solve this, is that there will at least be a fight about it." In our national politics, in our media, even in most of our environmental organizations, "we haven't even gotten to that point where there is at least a genuine fight about what, really, we are facing. That, to me, is the objective."

There's a kind of brutal logic, a strategic calculus, at work in what Ken is saying. It's a waste of time, "a fool's errand," he argued, for activists to obsess over the policy details of moderate, incremental approaches—a modest cap-and-trade program, a "dinky carbon tax," piecemeal EPA regulations—that are designed within the impossibly narrow constraints of our current politics. Because they won't come close to dealing with the problem—and because a farcical, gridlocked partisan debate over nonsolutions only distracts from any real reckoning.

What will that reckoning look like? "I do think we're in a period of collapse," Ken told me. "It's unavoidable. Even if we avoid the tipping points, we're going to reach a point of intense crisis, an emergency point, when the ordinary workings of government, the economy, and society are no longer possible. We don't know how soon that will be, but at that point, we'll be faced with some radical alternatives, some of which will be highly technocratic, highly dangerous." It will be geoengineering, he said, "like lofting a billion Mylar balloons into semistationary orbit"—or something more militaristic, "putting troops into the coal plants."

"Those will be our fundamental choices, within Eli's lifetime. There will be a series of radical responses on the table—and we need to win that fight. So as far as I'm concerned, everything we're doing should be building to that point, trying to create what we do not have, which is power, discipline, an idea, something to win that. Because otherwise we really will have the Mylar balloons."

I asked Ken what he really hoped to achieve with the lobster boat.

"I'd like to shut the plant down," he said.

Of course, he knew that a single act of nonviolent direct action, however dramatic, wasn't likely to accomplish that. Even a powerful and sustained grassroots effort would face a long, uphill fight. Which, to Ken, was precisely the point: the fight. Drawing the bright line.

Just suppose, I asked, that you did manage to shut the plant down immediately, however unrealistic that is—what would the impact be, economically and politically? Ken thought about it for a moment. "Compared to what?" he answered. "Compared to not shutting it down? Well, no, we can't *not* shut it down. So if the question is, 'What do we do after we shut it down tomorrow?'—I'm not qualified to answer that. Somebody else will have to figure that out. There's no question, though, that we *have* to shut it down."

Days later, putting their bodies and their freedom on the line to physically block the *Energy Enterprise* and its forty-thousand-ton cargo of coal, Ken and Jay dramatized the stark choices we face as a society and a species. They want us to understand: as human beings, we can have coal plants, or we can have a livable future. But we can't have both.

———

"This is about us, and our relationship to the planet, and our relationship to each other," Jay told me, a week before the action at Brayton Point.

We were sitting in the sunlit living room of a friend's apartment in Somerville, a block or two from Davis Square, but Jay lives down on Cape Cod, where he grew up, "just over the bridge in the town of Bourne."

Jay is slender, fit, medium height, with a dark trimmed beard and dark hair pulled back into a ponytail. He has a calm, resolute voice and a keen sense of humor. His upbringing, he told me, was "a very normal, middle-class life"—except that his parents, both public school teachers, had the summers off, and they would often sail with him and his younger sister to Maine.

"We spent a lot of time out of doors that way, being in the natural world," Jay said. "And I can remember, in high school, I knew that something was wrong with the orientation of human beings toward the world around us, and in relation to other human beings and to community."

Intent on keeping his livelihood separate from his activism, Jay was then working an hourly job as a sail maker to support himself, living lightly and simply—in the warm months, he lived on that same family sailboat he knew so well as a child—and devoting the rest of his time to climate organizing, primarily within the Quaker community.

Though his parents are not Quakers, Jay went to Earlham College, a small but prominent Quaker school in Indiana, where he became interested in Quakerism. "I was raised to be a good Congregationalist," Jay said. "We went to church every week." As soon as he got to Earlham, though, Jay started going to Quaker "meeting" (as Quakers call their worship services). The director of campus ministries asked him when he was going to become a member. "She was this awesome, radical black lesbian feminist from Philly," Jay recalls fondly. "I was like, 'Well, I'm not sure if I quite stack up to what Quakers should be, or could be, and I'm still trying to figure out what that is.' And she said, 'Jay, if you're asking yourself what it means to be a Quaker, you are one.'" He's now a member of Sandwich Monthly Meeting on Cape Cod—founded in 1656, the oldest continuous Quaker meeting in America—and attends the weekly West Falmouth Meeting, where he serves as Clerk of the Ministry and Counsel Committee, "tasked with pastoral care and spiritual nourishment."

Jay was a sophomore at Earlham on September 11, 2001, and as for so many people, the attacks and the US response changed everything for him. "It became very clear that I could no longer stay silent," Jay told me. "I turned into the consummate antiwar activist." At that point, he remembers, "the environmental concerns really didn't stack up, compared to the fact that we were killing people, *actively*, instead of passively."

He graduated in 2004 and went straight to work on John Kerry's presidential campaign in New Mexico, working with the League of Conservation Voters—"because that's what had to be done," he said. "We worked our asses off and lost."

Jay always had a strong interest in politics and policy, and he planned to go to Washington. But something pulled him in another direction. First, he told me, he needed to go for a long walk. He decided to hike the Appalachian Trail, from Georgia to Maine. "I spent four and a half months in the woods," he said, "walking." Anybody can do it, he assured me. "Grandmothers walk it, high school kids walk it, blind people walk it." And on the trail he discovered something profound—"intense community."

"It's people from all walks of life," he said, all political and religious persuasions—"evangelical Christians and backwoods Vermont hippies" (who are sometimes one and the same). "I met coked-out truck drivers and a past president of Jaguar USA. And all these people are together, all having the same intense experience, and they're looking out for each other, caring for each other, encouraging each other to find their own path, to do what they need to do to fulfill their journey. And it was so clear: this is how human relations are supposed to happen, we've created the 'blessed community' right here. I left the trail very clear that we needed to make the world more like the trail."

"The core of Quakerism," Jay told me, "is that you can't do it alone, that you have to do it in community, and that your own redemption comes in the redemption of the community."

So Jay moved to Washington and went to work for the Friends Committee on National Legislation, the major Quaker lobbying organization, focusing on peace and justice issues—"defund the Iraq war, repeal the Patriot Act, all those good lefty things." But working in DC, he began to see that "the problems of the world are too large to be solved from Washington, because those organizations operate within the constraints of what is 'possible' within Washington."

His disillusionment with Washington was complete—and his reckoning with climate began—when, in 2006, he read *New Yorker* writer Elizabeth Kolbert's *Field Notes from a Catastrophe: Man, Nature, and Climate Change*, which had just been published. "I remember lying in the grass outside Eastern Market on Capitol Hill, bawling my eyes out reading that book. That's what opened me to the climate crisis." Later that summer, he saw Al Gore's *An Inconvenient Truth*. "And that just drove it home for me. I thought, This is it. This is it. This is the thing that epitomizes our disconnect from each other and the world."

It's the master symptom? I asked.

"It's the motherlode of bullshit. If you can't solve this problem, you can't solve any of the others."

While in Washington, Jay had read a lot of Wendell Berry, who became a hero of his ("I named my cat Wendell"), as well as Bill McKibben, in particular his 2007 book, *Deep Economy*. Taking a cue from Berry, who as a young man left literary New York behind and returned to his ancestral Kentucky farmland, Jay moved back to Cape Cod. He was done with Washington and the inside game of politics.

"I made this very conscious decision, influenced by Berry, to move back to my hometown—knowing that if I was going to do this work, and make the claim that we have to radically change things, I can't go someplace that isn't mine to make that claim. I have to move to the place where I belong, to make the claim that we must do this together."

He sought out those who were doing grassroots climate organizing in New England, especially young people and students, and he soon came across a group in the Boston area that was putting together a large conference on climate in the spring of 2008. Calling itself Massachusetts Power Shift, it was modeled on the first national Power Shift conference the year before in Washington, which was attended by some six thousand students and young people representing an emerging youth climate movement. Jay e-mailed the Mass. Power Shift organizers, saying he'd noticed that they were having a lobby day on Beacon Hill, and that he'd just spent two years working for a lobbying organization in DC, so perhaps he could help out with that.

"And this guy Craig calls me up and he's like, 'Hey, Jay, great to meet you. Do you want to run the lobby day?'" Jay laughed. "And I'm like, 'OK.'"

That was Craig Altemose, then a student at the Harvard Law School and Kennedy School, who went on the following year to start the climate-justice network Students for a Just and Stable Future (SJSF). In

2010–2011, Craig cofounded Better Future Project (BFP) in Cambridge along with a community organizer named Marla Marcum, a United Methodist minister who had recently left her PhD program in theology at Boston University; and an organizer in Somerville named Vanessa Rule, a mother of two, who was engaged in grassroots climate and sustainability work. (In the spring of 2012, when BFP launched the 350 Massachusetts network, I had just joined BFP's volunteer board and I worked closely with Vanessa, who spearheaded 350Mass and was its lead organizer, and who went on to become cofounder and lead organizer of the climate group Mothers Out Front. The 350Mass network grew quickly and has made its presence felt in state politics, organizing around divestment, carbon-tax legislation, shutting down coal plants, opposition to new "fracked gas" infrastructure—and pressing the governor to fully enforce the state's pace-setting Global Warming Solutions Act.)

Jay dove in with Mass. Power Shift, and in 2009, together with Craig and Marla, launched Climate Summer—a program to develop student climate leaders, who spent the summer biking around New England in small teams, building relationships in the communities where they stayed, talking about the climate crisis, and gaining organizing skills (a number of Climate Summer alums have gone on to play important roles in the climate movement in New England and beyond). He worked nonstop on Climate Summer for a year and a half, and served as BFP's board chair, but soon found himself utterly burned out—"crisped," as he puts it. This wasn't the kind of work he was meant for, he realized. He knew that he needed to get back among the Quakers, where he belonged, and "align inwardly"—to go deeper into the Quaker community.

"I mean, if the *Quakers* aren't even doing this," Jay told me, "then we're *totally fucked.*"

I laughed, and so did he. Then Jay went on, his tone now deadly earnest.

"If we aren't willing to put our lives behind the words we speak, then those words are *worthless.* The world does not change *just because we say things.* Just like politics in Washington doesn't change when someone writes a very well-reasoned, perfectly footnoted argument about how we need to have a global climate policy. Because it doesn't have *power.*"

When Jay did finally return to DC, in August 2011, he was one of the 1,253 people from around the country—including his colleagues Craig, Marla, and Vanessa—who were arrested in front of the White House in that historic first civil-disobedience protest against the Keystone XL pipeline.

"We do have to put our skin in the game," Jay said. "If we believe certain things about how the world should be, and if we really believe they're true, they're only going to be made true in the world if we manifest them ourselves."

"This is why I think civil disobedience is so important," he said. It's not about getting arrested or challenging authority. It's about drawing a clear moral line. "Civil disobedience makes manifest the tensions that exist in society. It makes them real, in the world, so you can visibly see the tension between what is right and what is wrong."

I wondered if Jay ever worried (as I myself do) about coming across as "holier-than-thou." But before I could form the question, he beat me to the punch.

That dividing line, Jay said emphatically, "is *within us.*" We can't do this out of self-righteous anger, he said, "because we are all involved—it is *we* who are burning the coal, *we* who are creating the demand for tar sands." If we act in self-righteousness and anger, he said, "we tear ourselves to pieces, and we're useless. We have to do this out of *love.* 'Darkness cannot drive out darkness; only light can do that,'" Jay said, quoting Martin Luther King Jr. "It can only be driven out by love."

I knew that Jay resisted labels, but I had to ask him if he still considered himself an "environmentalist," or a "climate activist," or even a "radical."

"I'm a Quaker," he answered.

Do you have something against any of those other labels? I asked.

"I think those labels are *fucking bullshit.* I think it's ridiculous that we try to divide everyone up into a specific box that we can plug them into, so we can understand them—and marginalize them. I am so done with that. Because this is about humanity. This is about all of us, together, trying to understand and reconcile our differences with the laws of physics and chemistry. And we can't do that as a special interest group. The special interest is called *living*."

I asked him if putting his body in the way of a coal shipment is radical.

"I don't think it's radical," Jay replied. "I don't think wanting a livable planet is radical."

How far was he willing to go? I asked.

"As far as I'm led," Jay said. "As far as what is opened to me. I think there is an inner resource that all of us have, an inner guide, an inner wisdom—some people may call it God—that we can open our hearts to, that clarifies how we are to act in the world. We need to remain humble in our assessment of what we can actually do. I don't pretend to know what the five-year plan is for me."

What if the five-year plan was prison? Would he be OK with that?

"That's not the goal. But I could be open to that."

What about others? What about myself? I told Jay there was a part of me that was seriously tempted to get on that boat with him. Do all of us need to be open to that?

"We need to discard our preconceived notions of what we are 'supposed' to do, and figure out, when I sit by myself, on a mountaintop or next to the ocean, or in my living room, and I know that the world is such a way, and I know that the world needs to be such another way, am I able to live with myself and get up in the morning and act according to what I know is true? Have I done what needs to be done?"

————

In the late 1950s, at a time when Cold War dissent still marked one as dangerously fringe, an American Quaker named Albert Bigelow knew what needed to be done. The fifty-one-year-old painter and architect from Cos Cob, Connecticut, who had commanded US Navy combat vessels in World War II, was now a pacifist and student of Gandhian satyagraha. He had been one of eleven protesters, members of the ad hoc group Non-Violent Action Against Nuclear Weapons, who were arrested for civil disobedience at the gates of a nuclear test site in Nevada on August 6, 1957, the twelfth anniversary of the Hiroshima bombing. Charged with trespassing and given suspended sentences, the protesters received little attention. That September, when the United States government announced a new series of tests in the Pacific to take place the following April, Bigelow and others felt the time had come for more dramatic steps.

Inspired by the British Quaker Harold Steele, who in the spring of 1957 tried and failed in a similar venture, their idea was a "harmless, peaceful act," Bigelow would later write—"an act that could not be bypassed, could not be brushed aside, could not be ignored. An act that was a symbol. An act that would be a magnifying glass to focus the rays of conscience at the center of the problem—the nuclear tests."

On January 9, 1958, on letterhead of Non-Violent Action Against Nuclear Weapons, Bigelow and William Huntington, as captain and first mate of a thirty-foot sailing ketch called *Golden Rule*, along with another crew member, George Willoughby, informed the president of the United States and his administration of their intentions:

Dear President Eisenhower:

We write to tell you of our intended action regarding the announced spring test explosions of American nuclear weapons.

Four of us, with the support of many others, plan to sail a small vessel into the designated area in the Pacific by April 1st. We intend, come what may, to remain there during the test period, in an effort to halt what we feel is the monstrous delinquency of our government in continuing actions which threaten the well-being of all men. . . .

For years we have spoken and written of the suicidal military preparations of the Great Powers, but our voices have been lost in the massive effort of those responsible for preparing this country for war. We mean to speak now with the weight of our whole lives. . . .

In February, they set sail from San Pedro harbor in southern California, headed for the test zone near Eniwetok and Bikini atolls in the Marshall Islands, but were turned back by a fierce storm. On March 25, they sailed again. As Bigelow explains in his valuable 1959 account, *The Voyage of the Golden Rule*, the plan was to touch at Honolulu, 2,500 miles from San Pedro, before continuing the final 2,000 miles to their destination. But on April 11, the Eisenhower administration, through the US Atomic Energy Commission, issued a new regulation barring citizens from entering the test area. When the *Golden Rule* reached Oahu, its crew were placed under a restraining order, preventing them from leaving the dock. They repeatedly violated the order, in carefully considered acts of civil disobedience, and were arrested. Bigelow and Huntington, after a much-publicized trial, were eventually sentenced to sixty days in jail.

Somewhat miraculously another boat, the *Phoenix*, had arrived in Honolulu at around the same time, and its crew—a family of four—found themselves caught up in the *Golden Rule*'s story. The skipper of the *Phoenix*, an American anthropologist and yachtsman named Earle Reynolds, his wife, Barbara, and their teenage son and daughter, were nearing the end of a round-the-world voyage. Reynolds had worked in Hiroshima studying the effects of radiation on children, and though he and his family were neither Quakers nor seasoned peace activists, they were deeply moved by the courage and commitment of Bigelow and his mates. After anguished soul-searching and deliberation, the Reynolds family, together with a young man from Hiroshima named Nick Mickami, decided to proceed to the Marshall Islands test area in the *Golden Rule*'s stead. Slipping out of Honolulu, they made it into the test zone, but were quickly intercepted by the navy. Back in Hawaii, Earle Reynolds was tried and sentenced to six months.

The *Golden Rule*'s case had received widespread coverage, with thousands of supporters in the American peace movement cheering them on, but few found out about the *Phoenix*. Afterwards, Reynolds wrote to Bigelow: "*Phoenix*, in its trip, *was* the *Golden Rule*. I would be entirely happy if the entire world should think it was the *Golden Rule* which achieved its purpose, because it did!"

In "Why I Am Sailing into the Pacific Bomb-Test Area," published in the February 1958 issue of *Liberation* magazine, Bigelow described his family's experience hosting two "Hiroshima maidens" in their home, young women who had been disfigured by the atomic bomb and had been sent to the United States for reconstructive surgery. The connection Bigelow formed with them was transformative. "I am going," he wrote, "because, as Gandhi said, 'God sits in the man opposite me; therefore to injure him is to injure God himself.'

> I am going because. . . . without some such direct action, ordinary citizens lack the power any longer to be seen or heard by their government. . . .
>
> I am going because it is now the little children, and, most of all, the as yet unborn who are the front-line troops. It is my duty to stand between them and this horrible danger.
>
> I am going because it is cowardly and degrading for me to stand by any longer, to consent, and thus to collaborate in atrocities. . . .
>
> I am going because I have to—if I am to call myself a human being.

The voyages of the *Golden Rule* and *Phoenix* failed, of course, to change the course of US policy. But they were part of a mid-century peace and racial-justice movement that would have a profound influence on the antiwar and civil rights movements, as well as the environmental movement, of the sixties and seventies. One of the influential documents of that mid-century movement, cited by Bigelow and Huntington, was a pamphlet titled *Speak Truth to Power: A Quaker Search for an Alternative to Violence*, published in 1955 by the American Friends Service Committee, one of the most important Quaker organizations. Its authors, among them the civil rights leader (and Quaker) Bayard Rustin, held up Gandhi's example and argued not just for pacifism but for nonviolent resistance as the only adequate response to overwhelming power. Faced with the existential threat of nuclear war, they held up the force of individual conscience, pressing the case for fundamental social and economic reforms and the end of the US military-industrial establishment.

The closing pages of *Speak Truth to Power* still resonate, perhaps more than its authors could possibly have imagined. "Mankind, we believe, has a higher destiny than self-destruction," they wrote. "We have tried to face the hard facts. . . . How shall man be released from his besetting fears, and from his prevailing sense of futility? To risk all may be to gain all. We do not fear death, but we want to live and we want our children to live and fulfill their lives." Only "courageous non-violence," they concluded, "can overcome injustice, persecution, and tyranny."

Perhaps it's a stretch to compare the *Henry David T.* to the *Golden Rule* and the *Phoenix*, or to place Ken's and Jay's words alongside those of Albert Bigelow and the authors of *Speak Truth to Power*. Or maybe it's not. Maybe Ken and Jay, and the many others who have put their bodies and freedom on the line to disrupt and ultimately stop the extraction and burning of fossil fuels, deserve to be remembered for their part in the ongoing fight for human survival.

Indeed, it's worth pausing to consider the Cold War context in which Bigelow acted as it compares with our own historical moment. While the nuclear nightmare still haunts us, it's been joined and overtaken by an existential dread of climate chaos, civilizational collapse, and mass extinction. The military-industrial complex still reigns, but it appears almost benign compared to the science-denying fossil-fuel lobby and the planet-destroying carbon-industrial complex. (Bob Dylan's "masters of war," those who built "the big bombs," inflicted "the worst fear"—"the fear to bring children into this world." Today, we have our masters of warming.) And whereas *Speak Truth to Power* argued for a radical shift in foreign policy based on nonviolence, even the dismantling of the US military establishment, today's climate radicals—those honestly facing the hard facts—know that nothing short of putting the fossil-fuel industry out of business will suffice now. The latter may sound every bit as quixotic as the former—and yet, given the stakes, every bit as morally serious.

In the end, however, there's a desperate and all-important difference between the two historical moments that cannot be ignored: the Cold War deterrent of mutual assured destruction—however unconscionable—had its effect. Barring accident or suicidal madness (ever-present possibilities, then and now), the leaders of the nuclear powers simply had to refrain from pushing the button in order to avoid destroying the world. That didn't stop some of them from playing insane games of brinksmanship, but the nightmare of nuclear annihilation was and is, however precariously, preventable.

Today, the climate crisis has long since passed the point of prevention. Climate catastrophe, on some scale, is winging toward us: the missiles have left the silos, the bombs have left the bays. Cities, indeed entire countries, will be lost. Millions will needlessly suffer and die—just how many millions depends on how much more carbon is extracted and burned. In other words, Ken's and Jay's sense of urgency is *at least* as justified as Bigelow's and that of the authors of *Speak Truth to Power*. The task now, impossible as it may seem, is to prevent the entire carbon arsenal, or even any sizeable fraction of it, from being launched—and to salvage what we can of a livable planet.

———

As it turned out, Ken and Jay were not arrested that day at Brayton Point. They knew that they were taking a significant legal risk—that because they operated on the water, and because they were interfering with a power plant, a piece of vital energy infrastructure, they each risked federal charges that could have led to years in prison. But they also knew that they could be treated more leniently, and that appeared to be the case. Once it was clear that the anchor would be removed by a salvage crane that evening, the Coast Guard allowed them to motor the *Henry David T.* out of the inlet the same way they'd come in. But Ken and Jay were informed, there on the spot, that for obstructing a navigable waterway they were subject to a federal fine of $40,000 per day. "We were prepared to go to jail," Ken told me afterwards. "What we weren't prepared for was bankruptcy."

The authorities took a different tack, however. The federal fine was rescinded, and Jay, as skipper, was served with state charges including negligent operation of a vessel and failure to act to avoid collision, and both he and Ken were charged with disturbing the peace, disorderly conduct, and conspiracy. That last one, which carried a potential two-year jail sentence, was the most serious concern (and not just for Ken and Jay). But there would be no plea deals. They would go to trial, with jail terms and stiff fines a real possibility.

At the time of their action, a coalition of environmental groups called Coal Free Massachusetts—including among others Clean Water Action, Toxics Action Center, the Massachusetts Sierra Club, Conservation Law Foundation, and Better Future Project—had been working for years to close Brayton Point, by 2020, citing the health impacts of coal on the surrounding community and, to a lesser extent, the climate effects of coal as a major source of carbon emissions. When Ken approached the coalition about his and Jay's plans, some members discouraged him from carrying out the action, afraid it would become a political liability in the heavily working-class and economically struggling town of Somerset, which depends on the plant for a large share of its tax revenue. But Ken was adamant: nobody to his knowledge had ever demanded the immediate closure of a coal plant, purely on the basis of climate, and he and Jay were determined. They showed little interest in the familiar environmental-justice arguments for closing the plant, or in advocating for the state to ensure a "just transition" for the community, which they believed undercut the message of urgency. (On this, Ken and Jay differed with some members of their support team and many others within the 350Mass community. The tension between urgency and economic justice is not easily reconciled.)

That March, before the action, Brayton Point's owner, the Virginia-based Dominion, announced that the plant would be sold (at a fire-sale price) to the New Jersey–based private equity firm Energy Capital Partners. On the day of their action, Ken and Jay had a letter delivered to the heads of Dominion and ECP, cc'ing Massachusetts Governor Deval Patrick, calling on them to halt the sale and shutter the plant. The sale went through, but in the fall ECP announced it would close the plant in 2017. Analysts cited economic factors, primarily the low price of natural gas, which had already spelled the demise of the state's two other remaining coal-fired plants, in Salem and Holyoke.

But it's reasonable to think that more than simply economic factors were involved. That July, inspired in part by Ken's and Jay's action, some four hundred protesters from around the state, organized by students and grassroots climate activists with SJSF, Better Future Project, and 350 Massachusetts, marched on Brayton Point. Forty-four of them were arrested for peaceful civil disobedience at the gates of the plant, calling on Patrick to make Massachusetts coal-free and provide economic assistance to the community of Somerset. The announcement of the plant's 2017 closure, three years sooner than the environmental coalition had been calling for, was widely celebrated—but not by Ken and Jay, for whom 2017 wasn't nearly soon enough. It should have been closed years ago.

In January, Ken's and Jay's story took another unexpected turn. At their pretrial hearing in Fall River, both the judge and prosecutor agreed to let them proceed with a "necessity defense." Ken and Jay would not contest the facts of the case, but instead would argue that under the circumstances their actions were justified. If it went forward, it would be the first necessity defense in a civil-disobedience trial in the United States centered exclusively on climate.

The trial date was set for September 8, 2014. Among the expert witnesses who agreed to testify on Ken's and Jay's behalf were Bill McKibben and James Hansen.

I wondered if Ken thought to tell that psychiatrist in Hingham.

———

On a fine Sunday morning in June 2014, a year and a month after the Brayton Point blockade, Jay rowed me out to his family's thirty-four-foot sailboat, *Footloose*, at anchor in West Falmouth Harbor. We had just walked the half mile or so from West Falmouth Friends Meeting House, where I'd joined him for the traditional silent worship. I'd watched and listened as Jay rose during meeting, as Quakers do, to speak from his heart about the transformative power of love.

We went below deck, to the well-appointed cabin where he lives in the summer, and ate a simple lunch of sandwiches and greens Jay prepared from his plot in the community garden down the street. After we ate, we talked about what it means to be honest in the face of our crisis.

First, Jay said, the response needed is existential. "You have to let go of what was. The world you thought existed, the way you made sense of the world in the past, no longer applies." In that process of letting go, he said, "there's an enormous amount of grief and struggle. But that grieving is absolutely essential. In order to have the existential response, you have to have the existential crisis."

Second, he said, honesty means acknowledging "that there is nothing left, nothing else that we can hang our hat on, but each other. All these structures, capitalism and industrial civilization, are quicksand. And the only rock we have left is God and our neighbor."

I asked Jay if he was still working for social change, in the conventional sense, building a social and political movement.

"I don't think that gets deep enough for what I'm after," he said. "Social change almost always assumes the kind of political and economic structures that we have—where we tinker around with the pieces but the overall framework is set."

"I don't believe that," he went on. "I believe there are still powerful forces of history to be unleashed. I believe it's time for a new world—for the revealing of a new way of being in the world, of our relation to the earth and to one another."

What's going to unleash those powerful forces?

"Love."

And what's going to unleash the love? A movement? Climate chaos itself?

"I don't know," Jay said. What he does know is that in order to bring about the kind of monumental change we need, "we need people's whole hearts—and full commitment."

Jay looked at me from across the table in the tiny boat's cabin. "The love and commitment of communities of resistance, who are prophetic in their relationship to the existing world, who live into that world as faithfully as they can, not knowing where they're going to end up, but knowing that the root is love—that's the only force powerful enough to confront what we have to confront."

He reached over to the shelf behind him and handed me a book. It was a well-thumbed first edition of *The Voyage of the Golden Rule*.

FIVE

A Long-Haul Kind of Calling

And He lifted up his eyes on his disciples, and said, Blessed be ye poor: for yours is the kingdom of God.

Blessed are ye that hunger now: for ye shall be filled. Blessed are ye that weep now: for ye shall laugh. . . .

But woe unto you that are rich! for ye have received your consolation.

Woe unto you that are full! for ye shall hunger. Woe unto you that laugh now! for ye shall mourn and weep.

—THE GOSPEL ACCORDING TO SAINT LUKE (KING JAMES VERSION)

If I could do it, I'd do no writing at all here. It would be photographs; the rest would be fragments of cloth, bits of cotton, lumps of earth, records of speech, pieces of wood and iron, phials of odors, plates of food and of excrement. . . .

A piece of the body torn out by the roots might be more to the point.

—JAMES AGEE, *Let Us Now Praise Famous Men*, 1941

Early one morning in July 2013, I drove north out of Houston at the crack of dawn, three hours up Highway 59 into the cleaner air and the dense piney woods of deep East Texas. It was Sunday, and I was on my way to church.

I'd been up that way before. My father, son of sharecroppers and day laborers, was born and raised in northeast Texas, outside Paris, in the

Great Depression; he remembers his mother picking cotton, pulling him behind her on a sack, and his father's long absences, looking for work. And as I've said, my whole family is from Texas. Only the day before, I'd met up with an older cousin of mine in Victoria, a couple hours southwest of Houston, where my mother was born (though she grew up in West Texas, in the small town of Brownfield, south of Lubbock, and in Abilene, where she met my dad at the local Christian college). We had breakfast at a diner on that same Highway 59—eggs, sausage, and bacon, and most important, biscuits and gravy like my mom cooked on Saturday mornings in California—and then we drove out to the old farms and country churches where our grandparents and their parents had worked the land and prayed to God. We stopped and stood in front of the little white-clapboard Church of Christ, in a grove of big live oaks, where our forebears had opened their well-marked Bibles, sung hymns a cappella—same as I'd sung them as a child—and praised the Lord and judged their neighbors.

Suffice it to say, then, that I'm no stranger to Bible Belt Christianity. And yet I'd never been to a church like the one where I was headed that Sunday morning: the small, progressive Austin Heights Baptist Church in Nacogdoches, which meets in an unassuming building on the edge of town.

Austin Heights was formed as a breakaway congregation in the charged atmosphere of 1968, when its founders could no longer accept the dominant Southern Baptist line on issues of race and war, and it established a lasting fellowship with the major African American church in Nacogdoches, Zion Hill First Baptist. That first morning I was there, the Reverend Kyle Childress, Austin Heights' pastor since 1989 (and the only white member of the local black ministers' alliance), preached on the Hebrew prophet Amos—a prophet, he noted, who was among the favorites of the Reverend Dr. Martin Luther King Jr. Childress began his sermon by reminding us that 2013 marked the fiftieth anniversary of the protests in Bull Connor's Birmingham in the spring of 1963 and the March on Washington later that summer, and that one of King's most-used lines—found, for example, in his 1963 "Letter from a Birmingham Jail" and "I Have a Dream" speech—was a verse from that morning's scripture reading in the Book of Amos: "Let justice roll down like waters, and righteousness like an ever-flowing stream."

The prophet Amos, Childress told us, was called to be a fierce advocate—among the Bible's fiercest—on behalf of justice for the poor and oppressed. "Amos's strong preaching was hard then, and it's hard today," Childress said. Just as in Amos's day, when the wealthy trampled on the poor while worshiping piously in the temples, so today our "programs of care for the poor and needy" are dismantled "with a religious zeal." Meanwhile, "giant corporations get a free ride. They can diminish people, destroy the earth, pour out climate-changing carbon, all in the interest of short-term profit, and no one can do anything about it."

But Amos knew, Childress assured us, that God is the spring of justice—and without God, "we are unable to keep up the struggle for justice and goodness and love over the long haul."

Childress, himself a longtime activist on issues of equality, peace, and poverty—he founded and led the Nacogdoches Habitat for Humanity and the local AIDS agency—then offered a personal story: Twenty-five years earlier, he was working on homelessness in Atlanta with the Baptist Peace Fellowship of North America, and he and the other organizers had invited some African American clergy from the community to join their meetings. One day, he said, four or five ministers, young and old, "walked in wearing suits and carrying Bibles." Childress recognized one of the older men as the great civil rights leader C. T. Vivian, a friend and colleague of Dr. King's. Vivian, Childress reminded us, "had been beaten, jailed, knocked down, threatened, and almost killed, fighting for justice. C. T. Vivian was the real deal."

Out of respect, they asked if Vivian would lead an opening prayer. And so he put his Bible on his chair and knelt down over it, Childress recalled, "and he started praying, deep and hard." He thanked God for every detail of an ordinary day, and as he went on, Childress told us, "his voice picked up power and tempo, and toward the end of his prayer, with great conviction, his voice rose, 'Oh, God, send us the power of your Holy Spirit! You know the battle is hard and the journey is long! We can't make it without you!'"

"He knew what it took," Childress said.

"God calls us to justice, to be a people who embody justice," said Childress, his own voice gaining power. And yet, as King and so many others who fought for civil rights knew, "Walking with this God, knowing Jesus, living like Jesus and serving in the battle for justice, is a *long-haul* kind of calling."

Childress is not a Bible-thumper; he doesn't shout. Heavyset and ruddy faced, with a whitening, close-cropped beard, he was born and raised in Stamford, a small town north of Abilene, and speaks with a soft, flat West Texas accent. But his voice carries the strength of real conviction. Educated at Baylor and the Southern Baptist Theological Seminary, and deeply influenced not only by King and the black church tradition but by the Christian anarchism of Will D. Campbell and the Christian environmentalism of Wendell Berry, he wins you over with well-honed arguments and a down-home sense of humor. Which is to say, I would have been impressed with his sermon even if I hadn't known that his words that morning held a heightened significance for his congregation—not just because of the civil rights history, but because this little East Texas church, which could count perhaps a hundred souls in its pews on a typical Sunday, was involved in a new battle. I wasn't there just to hear the preaching.

The Keystone XL pipeline, specifically the southern leg of it, runs through East Texas, within twenty miles of Nacogdoches, carrying toxic and carbon-heavy Alberta tar-sands crude (in the form of diluted bitumen, or dilbit) from Cushing, Oklahoma, to the Gulf Coast refineries in Port Arthur and Houston for export. Soon after that southern leg was fast tracked by Barack Obama in March 2012, the Austin Heights congregation found itself in the thick of the grassroots battle to stop it. (According to TransCanada, the Canadian multinational that was building the pipeline, by the time of my visit that July, "Keystone South" was nearing completion. In January 2014, it went into operation).

I'd reached out to Childress after reading an essay of his titled "Protesters in the Pews" in the *Christian Century*, and following the morning service I was scheduled to meet and interview, there at the church, several members of Tar Sands Blockade (TSB). A year earlier, in the summer and fall of 2012, the diverse group of mostly young, radical climate- and social-justice activists, many of them Occupy veterans, had mounted a high-risk, headline-grabbing campaign of nonviolent direct action—including a dramatic, eighty-five-day aerial tree blockade and numerous lockdowns at construction sites—to stop or slow the pipeline's progress in Texas. In the process, they'd worked with everyone from local environmentalists raising the alarm on the dangers of tar-sands leaks and spills (as seen in Kalamazoo, Michigan, and Mayflower, Arkansas) to conservative landowners fighting TransCanada's use—and, they'll tell you, abuse—of eminent domain. Most of those who engaged in and supported the direct action campaign—a true grassroots uprising—were Texans, young and old, twentysomethings and grandparents.

That fall, a number of the young blockaders, looking to set up an encampment for fifty or more of them on private property just outside Nacogdoches, started coming to church at Austin Heights. And though they came from all over North America, and from all sorts of cultural and religious backgrounds, often with no religion at all, they soon formed a close bond with many members of the mostly white, middle-class congregation—who welcomed them into their homes like family. At the same time, they were working closely with the local grassroots antipipeline group NacSTOP—which stands for Nacogdoches Stop Tar-Sands Oil Permanently—cofounded in 2011 by several members of Austin Heights.

The morning after his sermon, I went over to Childress's house, a block or two from Stephen F. Austin State University (where his wife, Jane Webb Childress, is on the English faculty), and the pastor and I sat on the back porch drinking coffee. Ever since the blockaders began showing up at his church, he told me, people had noticed a change in his preaching.

"There's an urgency that maybe I didn't have before," Childress said. "They're reminding us that climate change is not something we're going to fiddle-faddle around with. I mean, you've got to step up *now*." The devastating Texas drought was all over the news, and he'd recently been to a funeral in central Texas, where a rancher from Llano told him they'd had no real rain since 2004. "That's nine years," Childress said.

But there's more to it, Childress told me. "I'm preaching to young people who are putting their lives on the line. They didn't come down here driving a Mercedes Benz, sitting around under a shade tree eating grapes.

They hitchhiked. They rode buses. And they get arrested, they get pepper sprayed, they get some stiff penalties thrown against them." In January, Tar Sands Blockade and allied groups had settled a lawsuit brought by TransCanada seeking $5 million in damages for construction delays, forcing them to stay off the pipeline easement and any TransCanada property. It was later revealed that TransCanada briefed law enforcement on the potential "terrorist" threat posed by the nonviolent protesters—and, as reported by the *Guardian,* that the FBI's Houston office spied on the group in Texas, violating its own guidelines for such sensitive political matters.

Childress noted that some of the blockaders, the Occupy veterans in particular, referred to the corporate-capitalist system killing the planet as The Machine. "And they're exactly right, using that kind of language," he said. "They're going up against The Machine in a real, clearly defined way. Not subtle—really upfront. And I'm trying to help them realize what it's going to take to sustain the struggle."

When it exploded onto the scene in 2012, Tar Sands Blockade galvanized a national climate movement that was ready for escalated direct action to stop the Keystone XL and build resistance to extreme fossil-fuel extraction: everything from the exploitation of tar sands, to shale oil and gas fracking, to mountaintop-removal coal mining. As several climate organizers engaged at the national level told me, the East Texas blockade showed the movement what it looks like to stand up and fight against seemingly insurmountable odds.

Now, a full year since it launched, and with the southern leg of the pipeline all but in the ground, I wanted to find out how—and even if—Tar Sands Blockade would go forward. In some ways, the challenges it faced reflected those facing the climate movement writ large. There was a tension, which many in the movement could feel, between the sheer urgency of climate action—the kind of urgency that leads one to blockade a pipeline or a coal freighter—and the slower, more patient work required for organizing and movement building over the long haul. I wanted to know what it would take for Tar Sands Blockade to sustain its struggle—not only what it took to get into the fight, in such dramatic fashion, but what it would take to stay in the fight.

That first Sunday at Austin Heights, I talked for several hours with four blockaders who were still living at the camp outside town. All of them had been arrested while participating in various actions on the southern pipeline route. One of them, a young woman in her early twenties who asked not to be identified, was a veteran of Occupy Denver who'd engaged in a high-risk tree-sit on the pipeline easement and whose legal case was still unresolved. Another, a recent MIT grad named Murtaza Nek, whose family is Pakistani American and whose Muslim faith is central to his climate-justice activism, told me he found a lot of common ground with Childress (or "Pastor Kyle," as they all called him) and the Austin Heights congregation. He'd been arrested back in January while serving as support for an action near Diboll, south of Nacogdoches. As it happens, I knew Murtaza from the Boston-area student movement, and I wondered how he'd been treated, as a dark-complexioned Muslim American. He smiled. "They thought I was Mexican."

A third blockader, forty-two-year-old Fitzgerald Scott, also an Occupy veteran—Tampa, DC, Denver—was a former marine who was born in Trinidad, grew up in Newark and East Orange, New Jersey, and had a master's degree in urban planning from the University of Illinois at Chicago. Fitz, as he was called, was the only African American blockader I met in Texas. He'd recently been arrested, not once but twice, for locking down at Keystone construction sites in Oklahoma with the Great Plains Tar Sands Resistance. He told me that he'd joined TSB out of solidarity with other activists and with people in frontline communities fighting the industry, not out of any deeply held environmental convictions. In his experience, he said, "the environmental movement was far removed from blacks. That was just a place we didn't travel." Now he was trying to organize among African American churches, starting with Zion Hill there in Nacogdoches.

The fourth blockader at the church that Sunday was twenty-two-year-old Matt Almonte. Back in December, he and another activist named Glen Collins locked their arms to one another and to concrete-filled barrels inside part of the pipeline that was under construction—and came close to being gruesomely injured, possibly killed, when police used heavy machinery to pull the pipe sections apart by force. Though he was charged with misdemeanors, Matt's bail was set at $65,000, and he spent a month in jail.

Matt, whose family is Puerto Rican and Dominican, grew up in urban New Jersey, "somewhere in the middle of poor and working-class," he said, before his parents made the "climb to affluence" and moved to "a gated suburban thing in Florida." He attended the University of South Florida, where he studied political science, but dropped out after three years, attracted to Occupy Tampa. At a Greenpeace direct-action training camp in south Florida, he met a young organizer from Texas, involved with TSB, who told him about Keystone XL. "She said there's this pipeline coming through, and when it does I'm going to call you," he told me. She kept her word, and he headed to East Texas in mid-November 2012.

Although Matt said he felt "an enormous spiritual tug toward resistance," he considered himself an atheist, and he confessed that he was highly skeptical about his first visit to Austin Heights. "Basically, I came here laughing," he said. "I didn't have a whole lot of confidence in my ability to assimilate. I'd heard there was this Baptist church in town that totally loves us, and I was like, well, we'll see."

"And then I get here," he said, "and I meet these people, and it's not what I expected at all. It's so warm and welcoming. And I continued to come every Sunday after that. We always loved the sermons."

For Matt, the lasting impact of Tar Sands Blockade "was to show 'ordinary people' that it's absolutely vital to take direct action, and that even in a community like East Texas, people are rising against the fossil-fuel industry." He emphasized that trainings and actions were being networked out across the country, in South Dakota, Oklahoma, and elsewhere along the northern pipeline route and beyond, "places that don't typically see a lot of environmental resistance"—like Michigan, where Michigan Coalition Against Tar Sands (MI-CATS) was fighting the extension of the Enbridge pipeline that spilled disastrously into the Kalamazoo River.

Matt told me he identifies first and foremost with anarchism, not environmentalism, and with the way "different forms of oppression and exploitation intersect with each other." The Keystone XL pipeline "isn't just an environmental issue, it's a human issue, a social issue." He called himself a radical, with intention, because "political systems have a vested interest in demonizing language—they demonize the word radical, or militant, or anarchist."

Matt seemed impatient for more escalated direct action, the kind that was no longer happening along the southern pipeline route. Shortly after we talked, he and the young woman at the church decided to move on.

In fact, by the time I arrived in Nacogdoches, the blockaders' numbers had dwindled—some who had come from out of state had returned home or drifted off to join other direct-action campaigns against Keystone and various extraction projects—and the group was at something of a crossroads. Indeed, they were wrestling not only with tactics and strategy but with the very nature of the campaign, now that there was essentially nothing left, in the near term, to blockade.

But a solid core of about twenty organizers, many of them young Texans, had regrouped in Houston and were attempting to shift into something more like community organizing, engaging with environmental-justice efforts on the city's hard-hit and largely Latino east side, along the Houston Ship Channel and the refineries at the terminus of Keystone XL. As several of them told me, they wanted to extend their campaign beyond the fight against Keystone and tar sands, and to build a base of grassroots resistance to the fossil-fuel industry right there in Texas—especially in frontline communities, often communities of color, most affected by fossil-fuel pollution. The kinds of places, they pointed out, where the climate movement had established little, if any, foothold.

When I got back to Houston, I sat down one morning with Ethan Nuss, Kim Huynh, and Ron Seifert, three members of TSB who'd taken on the role of spokespersons, and they talked openly with me about their campaign at that pivotal moment.

Ethan, who was twenty-nine at the time, with long brown hair and a boyish face, had been my first contact with TSB before arriving in Houston.

The oldest of five kids, he grew up in Salina, Kansas, "very much middle America," he said. "You know, hang out at the truck stop and eat French fries." His mother was an artist and his father was a lawyer and ex-marine from a long line of military men. Ethan said that growing up he wanted nothing more than to join the US Marine Corps. But by the time he got to college at the University of Kansas, in the aftermath of 9/11 and the march to war in Iraq, he did a complete 180 and became a campus antiwar activist. That led him to focus on America's addiction to oil, and from there to climate change.

Ethan had used the phrase "climate justice," and I asked him what he meant by that. "Climate justice is the acknowledgment of the disproportionate impacts of climate change on low-income communities and communities of color, and elevating their struggle, acting in solidarity with them."

Could it apply as much to a midwestern farming community hammered by drought, I wondered, as to communities in Houston hammered by pollution?

"I think so," he said. "I feel like our work has been most effective when we've worked directly with people who are impacted, whether it's the farmer in East Texas or our Indigenous allies in South Dakota, or the communities in Houston at the terminus of the pipeline, breathing the worst of the toxins."

Kim Huynh, then twenty-six, sat next to Ethan. She was born in a Vietnamese refugee camp in Indonesia and immigrated with her family, first to Canada and then Florida, where she went to the University of Florida and studied political science and sociology. Kim told me she came to climate activism "in a very academic way, through the intersection of things I'm passionate about: human rights, migrant rights, racial and economic justice." A year before, she'd left a job with Friends of the Earth in Washington, DC, focusing on Keystone and climate, and came to Texas to join the blockade. Working for FOE for over a year, she'd sat at the table with groups "who have institutional access and resources when it comes to the KXL fight," but she was "frustrated by this inward-looking echo chamber that is the Beltway." She had been attracted to direct action for a long time, "as an ethic, a philosophy, how one lives one's life." And she'd seen collective power at work in other movements, like the housing rights movement in DC, which "grew out of Occupy and had direct action at its core, as a value."

I asked Kim if she had any trouble reconciling the urgency of climate action, what drew her to the pipeline fight, with the kind of long-term commitment required for the movement building she wanted to engage in now.

"I certainly feel that tension," she told me. "A lot of folks that I've worked with feel that tension very strongly, feel it in their bodies. It's an anxiety." At the same time, she said, she also felt "a commitment to the idea that we need systemic change, like actually hacking at the roots of what climate change is and what's created climate change." That kind of change is a long-term thing, she acknowledged. "It isn't going to come just from stopping the pipeline. Stopping the pipeline is a good start."

"The challenge and struggle for TSB," Kim said, "is to figure out how we define escalation, as a campaign that started from this extremely escalated place."

She then drew a striking, and bold, historical comparison.

"Personally," she told me, "I draw a lot of inspiration and lessons from the black freedom movement, the civil rights movement, thinking about groups like SNCC and the way they defined escalation as going into the most deeply segregated areas in the South and doing voter registration." That's a whole other kind of escalation, Kim said, "doing the organizing in the areas where it's possibly most important to do. Maybe that strategy is less like direct action as we know it, lockdowns, and more like community organizing. But that doesn't mean it's any de-escalation."

"The communities that are most impacted by these industries," she said, "the people who are living and breathing it every day—they need to be leading the fight."

As Ethan and Kim both suggested, that idea of disproportionate impact—of both climate change and fossil-fuel pollution on hard-pressed communities that can least afford it—is at the heart of what Tar Sands Blockade meant by climate justice. They wanted a radical movement, one that would grasp the problem whole, at the roots of the system, and fight alongside those who were already on the front lines—and always had been.

~

When I first met Ron Seifert, we were standing outside on a sweltering early evening at Hartman Park in Houston's Manchester neighborhood, just east of the 610 Loop along the Houston Ship Channel and across the street from a massive Valero refinery. Ron was a founding member of Tar Sands Blockade—and, at thirty-two, was also among the oldest, with the first early flecks of gray showing in his trim black beard. Having trained for years in long-distance endurance racing, his slender frame seemed to conceal a reservoir of stamina.

Ron grew up in Wisconsin and South Carolina and came to Texas in late 2011 from Montana, where he'd been exploring grad school in environmental science and law at the University of Montana. While there, he'd had long conversations with a political science professor who wrote his dissertation on the Zapatistas uprising in Chiapas, Mexico, and Ron started thinking about radical social movements. Around that same time, he learned about the Alberta tar sands when he noticed trucks carrying massive industrial mining equipment, known as "megaloads," northward on the precarious mountain roads near where he lived. Like Ethan, Ron joined the historic sit-ins at the White House in August 2011. Later that fall, along with another activist named Tom Weis, Ron biked the full length of the pipeline route, from Montana to Texas. In the spring and summer of 2012, after Obama fast tracked the southern leg, he helped launch Tar Sands Blockade, together with members of Rising Tide North Texas, on landowner David Daniel's property near Winnsboro in the northeast part of the state, site of the storied eighty-five-day tree blockade.

Rural and small-town East Texas is a world away from Houston's Manchester neighborhood. Overwhelmingly Latino, the community is surrounded by oil refineries and other heavily polluting industrial facilities—a chemical plant, a tire plant, a metal-crushing facility, a train yard and a sewage treatment plant—and sits at the intersection of two major expressways. The people who live there already breathe some of the country's most toxic air, and they have the health statistics to show for it. Not just asthma and other respiratory problems—a recent investigation by researchers at the University of Texas School of Public Health had found that children living within two miles of the Ship Channel have a 56 percent higher risk of acute lymphocytic leukemia than those living only ten miles away. The Ship Channel and nearby refineries—along with the refineries near the struggling African American community in Port Arthur—were also a prime destination for tar-sands crude, only increasing the toxic emissions in these fence-line neighborhoods.

I was there in Manchester that evening to tag along with members of Tar Sands Blockade as they canvassed the community door to door, conducting a health survey in collaboration with the local Houston group TEJAS—Texas Environmental Justice Advocacy Services. We were also letting residents know about the upcoming Healthy Manchester Festival there at the park, where an alliance of groups representing labor, immigrant rights, and environmental justice would come together for a family-friendly afternoon of local food, music, and political speeches.

Later, TEJAS cofounders Juan Parras, a longtime labor and environmental-justice organizer, and his son Bryan, talked with me about the challenges the climate movement faces in places like Manchester, or any place where immediate health, economic, and social pressures are paramount. Broadly speaking, Bryan Parras told me, most efforts at climate action "tend to leave the same folks that are already in bad situations in bad situations. There's no incentive for them to get involved." When I talked with Bob Bullard of Texas Southern University later that summer, he echoed Juan and Bryan Parras. "We could stop every pipeline being built from Canada to Texas, we could stop every fracking operation, and still not deal with the justice question, what happens in these communities," he told me. "What we're trying to drive home with our friends and colleagues in our larger environmental movement, our larger climate movement, is to talk about these communities that are at greatest risk, put a real face on this. Make it real."

Tar Sands Blockade was listening. "Disproportionate impact is very real," Ron told me. He said that TSB wanted to support and amplify the work of TEJAS and other environmental-justice groups in Texas. At the same time, he and his TSB colleagues were conscious of what might be called the "parachuter syndrome," in which outside activist groups, however well intentioned, drop in uninvited and pursue their own agendas. Given that apparent tension, I asked Ron if there's not a disconnect of sorts between the kind of urgent climate action TSB embodied (literally) in its direct action campaign and the slower, and in some ways more difficult, work of environmental-justice organizing in these communities.

Maybe, he said. But, ultimately, "it doesn't matter."

"Pipelines and refineries and droughts—those are not different things," Ron continued. "That *is* climate change. The refineries are climate change. Keystone XL is climate change. Tar-sands exploitation is climate change. It's all the same thing. And we understand that these communities are bearing the brunt of this industry—which is one and the same as climate change. It's in their backyards. They have children in the neighborhood die. They understand this industry will kill you for profits."

I talked with Ron several times, and he told me that a core group of TSB organizers wanted to do the kind of climate-justice organizing that national environmental groups weren't doing in Texas. From the start, Ron said, Tar Sands Blockade had shown a willingness to defy a status quo within the larger movement, in which only "winnable" campaigns—such as stopping the northern leg of the KXL pipeline, but not the southern—were taken on and funded. With the fight in East Texas, and by digging in now for the longer, even harder fight in Houston, "we've been able to say this is worth fighting no matter what, even if it looks like we can't win."

"That type of real investment and commitment," Ron said, "the idea that you have to go into where the problem is worst—like Mississippi during the civil rights era—you have to get in there and get a foothold. We hope we can empower local-led action and resistance. In Houston itself, there are literally millions of people who are being poisoned. We should be able to empower folks here to rise up and defend their own homes."

In social movement terms, it sounded like Ron was sketching a broader theory of change. "If the climate movement is ever going to win in a really robust way, it's gotta come to Texas, the belly of the beast,"

Ron said. "Houston, and the Texas Gulf, is the lion's den—the largest petrochemical complex on planet Earth. If the base isn't there, if the communities there aren't organized and informed, empowered to take action, the movement isn't going to be successful when it needs to be." They may have failed to stop the construction of the southern pipeline, "but we can still build and cultivate a culture of resistance and action, capable of escalating to the point of shutting this stuff down in the future."

A few weeks later, after I'd left Texas, Ron wrote to me in an e-mail, trying to answer for himself the question, as he phrased it, "Not why do we get involved, but why do we persist?" He wrote: "Is our campaign 'unwinnable' from a mainstream NGO perspective? Yes." And yet, he told me, "Resistance has intrinsic value that exists regardless of its demonstrable efficacy. It's not just an outcome, it's a life's work."

———

Bryan Parras, with his short, powerful frame, black hair pulled tight in a ponytail, was thirty-seven years old when I met him that July in Houston, his hometown. He'd been working on environmental-justice issues there and along the Gulf Coast since 1999, when he graduated from the University of Texas at Austin, and together with his father, Juan, cofounded and built TEJAS, their small grassroots organization, on Houston's east side. Starting in 2005, working on the ground in New Orleans and the Gulf Coast after Hurricane Katrina, he got "a crash course, like a PhD," he told me, "in justice-movement organizing—on housing, immigration, worker rights," and he became an advisor to the Gulf Coast Fund for Community Renewal and Ecological Health. He had just recently completed a job with the University of Texas Medical Branch as lead field coordinator on a research project looking at the impacts of the 2010 BP drilling disaster on Gulf seafood and the communities that depend on it.

Gregarious, quick to laugh, and with a keen love of literature and history, Bryan had spent a lot of time doing what he called "cultural organizing," and had coproduced a radio show, *Nuestra Palabra: Latino Writers Having Their Say*, on the local Pacifica FM station. He told me that he identified with the Chicano movement, and that his environmental ethic springs from strongly held Indigenous values, a deep respect for the interconnection of the earth and all living things. Those values, call them spiritual, "have been taught out of us," he said, by industrial civilization. Asked if he considers himself Indigenous, he said, "Sure. Chicano. But I don't have a rez." Then, with a laugh, "The east side is my rez."

On top of his Houston and Gulf Coast work, Bryan had become more involved in the fight against tar sands and Keystone XL, in solidarity with the Idle No More movement connecting Native American groups and the First Nations peoples of Alberta. (Eriel Deranger of the Athabasca Chipewyan First Nation has called the ecologically and culturally devastating tar-sands operations a form of "cultural genocide.")

For some reason, I asked Bryan if he thought of himself as an environmentalist. "I let folks call me that," he said. "And I'll use that term if it's the only way of describing myself to someone that makes sense. That's the problem: We need more words. We need more words."

We were sitting outside at a coffeehouse that Bryan had suggested on Westheimer Road, in the gentrified Montrose neighborhood. It was a long way from Manchester, though with construction going on across the street, and diesel fumes drifting over the patio where we sat, the air wasn't exactly pristine.

"That's 'sweet' diesel," Bryan said.

We started talking about what I was writing and why I was there. I mentioned the tension I'd found, especially among climate-justice activists, between the urgency of the climate crisis and the need for real community organizing and movement building. I said that I was looking for "climate justice" in Houston and East Texas. It seemed to mean different things to different people, and I wondered what it meant to him.

"Nothing," he said.

Nothing? I asked.

"I know what people have told me it means. For me, and for others, it's about the disparities in who is impacted by climate change—the Global South, people of color, poor folks, fishing communities, subsistence farmers."

And future generations? I asked.

"Yeah. And all those are Indigenous concepts. Your duty to protect future generations, that's an Indigenous concept, and I think a very human concept, too. It's a religious concept. All this stuff is inherent in us."

So, why did he say climate justice means nothing, especially given where he lives and the work he's doing?

"It's too big," he said. "No one has done a good job of painting a picture of what it could look like, or should look like. It's sort of an ideology that has to be instilled in you. That's why I think there are very, very big cultural differences in how we look at something like this."

But he also pointed to overriding economic factors in addition to the cultural. "I have to have a job, I need to make a living, put food on the table. People have kids, they need clothes. Until we can really answer that, there's no reason for them to care about what's impending on us all, the climate disaster."

Most important, he said, for any organizers coming into a community like Manchester, is to see that it's a "dynamic situation." In other words, he said, "For the people there, a lot of the injustices are not only environmental, in the sense that most people think of the environment. So, chemicals, yes. But police brutality? Drugs, other kinds of chemicals? Lack of educational opportunities? And then on top of all that you have these environmental, toxic exposures."

What did that mean for climate-justice activists like TSB, who see their work intersecting with his? How could they actually help TEJAS in its work?

"Honestly," he said, "I think it's more important for *us* to help *them*. That's how I see my role. We all need to help each other. The reason these communities like Manchester exist is because there are deficits in other communities that are not paying attention to these neighborhoods and these people."

That made sense to me—certainly, climate-justice activists, whether TSB or someone like myself, have much to learn from the likes of Bryan and Juan Parras—but I still wasn't sure how it addressed that central tension around the urgency of climate action. Considering what's in store for places like Houston and the Gulf Coast, then what about the children of Manchester who will be alive a mere forty or fifty years from now?

"I don't know what the answer is," Bryan said. "But when you have these larger, global concepts, like 'climate action,' they tend to favor certain communities over others. If there were an honest concern about something as big as climate change, then we would be having discussions about more than just how much carbon is in the atmosphere."

———

Yudith Nieto is an organizer with TEJAS. We met in Manchester's Hartman Park on another hot and soupy July afternoon, and sat at one of the picnic tables under some shade trees by the dusty baseball diamond. Through the chain-link backstop behind home plate, it looked as though a deep shot into left field would bounce off the refinery tanks across the street. Just over the trees out in far center, a towering smokestack sent a gray plume into the heavy air. An acrid odor wafted over the park. All around us were the small houses and yards of the neighborhood Yudith called home.

Yudith had agreed to talk with me that afternoon before meeting up with Juan Parras and researchers from the Environmental Integrity Project, who would show her how to change the filters in an air-quality monitor they'd set up in a yard on the edge of the neighborhood, adjacent to Interstate 610 and the metal-crushing plant.

"People hate it," Yudith said, referring to the crushing facility. "They're tired of it—the noise, the dust particles, everything. They're literally sick." The company had built a wall, she said, "a fifty-foot wall, that's supposed to help with the dust. The logic was the particles respect walls and boundaries. You know, America's very fond of walls."

Yudith, who stands all of four foot eleven and has the face of an angel with dark brown eyes, was born in Mexico. "Reynosa, I think. One of the border towns," she said. "I was brought here when I was five years old. My family is originally from San Luis Potosí, in the heart of Mexico. It's a ways off."

"I'm twenty-four years old," she said. She became a naturalized US citizen when she was twelve.

I asked if she remembered leaving Mexico.

"Yes. It was very, very sad and frightening, because I was leaving all my friends and some of my family there. And then just being in a different place where you don't understand what anyone's saying. It was scary. It was completely different here, everything's paved, the skyscrapers were so high."

She said her family settled in Manchester. Her father did manual labor, and her mother stayed at home. "I started school in the first grade," Yudith said. "I didn't speak a word of English. Didn't know what was going on. It took me a while to get used to the refineries and the smells, because I didn't come from a place where they had refineries. So at first, you know, it was a weird experience for me, just the smells alone were so different."

I mentioned the odor I'd noticed when driving into the neighborhood.

"I think outsiders can smell it differently," she said. "I can't smell certain smells anymore, my nose doesn't work the same."

Her family had been hoping to relocate for years, ever since some other families had their property bought by Valero and were able to move out of the neighborhood. Yudith's family had a very different experience. When Valero bought the lot where they were renting, they were given a week to get out. But her parents asked the owner what would happen to the house, and they were told that nobody wanted it.

"They said, 'We're going to demolish it,'" Yudith told me. "'You can take the house. Just fill out this one-dollar form.' So we got the house and relocated it down the street, where we bought the lot. Lots and houses here are very cheap. We didn't go very far. My family couldn't afford to leave, to go somewhere else. So that's where the house is now."

Yudith noted that Valero's property has expanded significantly in the years she's lived in Manchester. "That house right there, surrounded by the tanks," she said—pointing north across the ball field to a small home with enormous industrial structures towering over it on three sides— "that was not like that, those tanks were not there when I was growing up. There were a lot of houses there. And over time, the industry grew on top of us."

"There's no zoning laws in Houston," Yudith pointed out.

I asked about her family's health, given the known toxics and human carcinogens in the air. "A lot of my cousins, and myself, we grew up with asthma and respiratory problems. You know, respiratory infections that last for months, that we can't get rid of, because we're constantly breathing this in. My grandmother gets really sick and we have to take her to the hospital because she can't breathe."

After high school, Yudith went off to the Art Institute of Houston, and as she spent less time in her old neighborhood she began to notice something. "I told my family, I wasn't getting sick a lot." She left the art school after two years, because she realized that the kind of corporate design jobs that would be available to her—most likely at one of Houston's ubiquitous oil and gas firms—wasn't what she wanted.

It wasn't long before she met Juan and Bryan Parras and got involved with TEJAS. In September 2011, TEJAS organized two events at Hartman Park to rally the Manchester community against Keystone XL, inform residents of the dangerous emissions from processing tar-sands oil at refineries along the Ship Channel, and to recruit people to testify at an upcoming public hearing on the pipeline in Port Arthur.

"TEJAS came into the neighborhood banging drums," Yudith remembered. "And some of the Free Radicals were there, this awesome Houston band that I love, some were really good friends. That was why I went to the park, because I saw my friends and heard the music. And there was Juan and Bryan, and all these people I'd never seen before in the neighborhood, talking about this pipeline coming in." She started asking questions, and Juan told her that they were heading the next day to a community hearing in Port Arthur. Yudith decided to go with them. The experience changed her life.

"There was a bus of ten or fifteen of us, and we get there, and the room is full of industry people," Yudith recalled. Many of them wore T-shirts with corporate logos. "It was really eye opening, and painful," she said. "They were hateful, snickering and making jokes at us. The most racist people I've ever met in my life. It was just disgusting. It was culture shock."

"I took it very personally," Yudith said. "I was trying to represent my family, people I care about, and they had no compassion toward people who would be suffering. All they talked about was money, jobs, 'what America needs.' No, this is about health, communities. I was more hurt than angry. I looked them in the eye and said, I don't want this in my neighborhood. How are we going to survive this? We already have enough shit going on in the air. We don't need any more. I couldn't handle how they were ridiculing what I was saying."

It seems that in that searing moment Yudith found the courage to stand up—and keep speaking. With the national Keystone fight intensifying over the next two years, she would soon find herself, in her work with TEJAS and Tar Sands Blockade, a rising voice for frontline communities in the climate-justice movement, invited to speak at conferences and on campuses around the country. She always traced it back to that hearing in Port Arthur.

"It helped me realize that nobody was speaking for my community," she later told me. "I had to reflect on what was important in my life, and the life of my family. Is it going to college and acquiring a degree and going to work for one of these oil companies? Is it just getting money and a house and a car like everyone else? I really had to dig deep, to what my priorities really were, and what did I really want to do with my life, with my voice, my conscience."

The following April, I went back to Houston, and I had a chance to catch up with Yudith again. Still involved with TEJAS, she was taking classes along with Bryan Parras to become a certified community health worker, both for some income—she could use the job—and to strengthen her environmental-justice organizing.

Things had quieted down since the height of the Tar Sands Blockade campaign. Keystone South had been operational since January, and while TSB had drawn attention to evidence of shoddy construction by TransCanada—and filed a Freedom of Information Act request that revealed sloppy inspection by PHMSA, the federal agency responsible for pipeline safety—many of the blockaders in Houston had continued to scatter. Others were focused on supporting tar-sands resistance elsewhere, from South Dakota to Maine to Mobile, Alabama—where TSB's Ramsey Sprague was organizing with Mobile Environmental Justice Action Coalition to oppose a dramatic expansion of tar-sands crude coming through for export. But a few of the blockaders in Houston were still actively working with TEJAS. I was curious to hear from Yudith how that relationship had been during the past year and more.

Early in 2013, with the TransCanada lawsuit deterring further blockades along the pipeline easement, TSB had shifted its direct action protests to corporate targets in Houston—disrupting oil industry meetings, occupying the TransCanada office downtown, even crashing a Valero-sponsored golf tournament. (This was during a period of intense activism nationally against Keystone XL and the tar sands, with TSB-inspired solidarity actions taking place all across the country.) Yudith was involved with some of the actions—when TSB's Perry Graham scaled a flagpole outside LyondellBassel's downtown Houston office, and secured a large banner protesting the company's plan to triple its tar-sands refining capacity at its Manchester facility, Yudith was one of the organizers and served as photographer and spokesperson.

But earlier, back in the late fall, TSB had gotten off to a rocky start in the neighborhood. In the first TSB-organized action in Manchester, two longtime environmental activists locked themselves to oil tanker trucks outside the Valero refinery and began a hunger strike, demanding that Valero divest from Keystone XL and put the money into improving Manchester's health and well-being.

"That was a problematic action," Yudith told me. "Some community folks appreciated that somebody was stepping up for them. But afterwards, some people were really upset because the TSB crew and other activists came into this community, planned the action, got media attention, but did not include the community that they were speaking for in the process. And you cannot speak for a community if they do not want you to speak for them."

Yudith said some feared greater surveillance and an increased police presence after that action. When Yudith would mention Tar Sands Blockade to people in the community, some would step back. "They'd get upset with me, like, 'Oh, you're one of them.' And I'm like, 'No, no, I live here.' I had to gain their trust."

In February 2013, Yudith traveled to Washington, DC, along with Tar Sands Blockade, to join a high-profile civil-disobedience protest against Keystone XL at the gates of the White House with Bill McKibben, the Sierra Club's Michael Brune, and forty-five others, including such notables as Julian Bond and Robert F. Kennedy Jr. That experience provided another lesson. Not only did she fail to gain any media attention for Manchester and TEJAS, despite the efforts of TSB, but when she returned home she had to face the reactions of family and friends. "Getting arrested—in the Latino community, Mexican American community, we don't do that," she said. "We do not get arrested on purpose. That is a historical trauma that we try to avoid. My whole family was scared for me, they didn't know where I was, they didn't know if I was going to be put in prison. Some even thought I was going to be deported."

TSB's organizational ethos and style of operating didn't help either. With many Occupiers among them, the group was governed by an anarchist-influenced, horizontal (as in, antihierarchical), and radically democratic "leaderless" decision-making process (which at least some of the TSBers I spoke with found frustrating and at times counterproductive). On top of that, the blockaders were still using their "tree names," the aliases they'd assumed for security purposes while in East Texas. This added a layer of secrecy and, ultimately, distrust to their relationship with TEJAS.

"We had all these questions," Yudith told me. "Who's making decisions? Who are these people? What are their real names? And who's their leader? Who sent them here? And they were never transparent about it. At one point, I really didn't want to have anything to do with them anymore. I was scared to use my phone. I was scared to talk to anyone. I started getting suspicious of them. And to this day, I still don't completely trust them. It's so unfortunate, because I do believe they're awesome people. But maybe if you want to help people, you should be open and sincere."

The working relationship improved, but not before some frank talk, followed by a decision that TSBers who wanted to help in Manchester would become TEJAS "interns."

"We had to give them that term, because they were calling themselves organizers," Yudith recalled. "And they were giving us agendas, how we needed to conduct our meetings. So we had to say, 'Hey, you're coming into our hood, we're going to tell you how we run things. You're going to sit there, and you're going to listen to us.' And we said it in the most polite, endearing way. But hell, in my head, I'm like, 'What the fuck are these people thinking? They're coming into my house, my territory, my environment, and telling me how I need to organize myself and my people? What?'"

In November 2013, I saw Yudith speak to a crowd of several thousand college students and young climate activists at the Power Shift conference in Pittsburgh. Despite her small physical stature, her presence on the stage in the cavernous convention hall was awe inspiring, a thing to behold. She held the room rapt as she told her story, utterly poised and direct in her plainspoken eloquence.

Yudith told me in Houston that when she started with TEJAS, she had had no plan to keep on speaking in front of crowds. "I failed my public speaking class! I've always been scared or embarrassed." She said that she still sometimes suffered anxiety and panic attacks. But she kept finding herself in those positions, she said, put there by "the Creator" or "Great Spirit." Before she speaks, she told me, she prays.

"I don't know what I'm going to say," she said. "I just pray. I ask for guidance."

Although she doesn't consider herself religious in any orthodox way, Yudith told me she has "an ingrained kind of spirituality." Her family is Catholic, she said, "but we're also part Nahuatl, which is Indigenous from central Mexico." A sincere spiritual seeker, she had even experimented with Zen and Tibetan Buddhism. "I used to go to a Tibetan

Buddhist temple for meditation and dharma talks," she told me. When I asked if she feels called to do the work she's doing, she said, "I can't explain it any way else. I did not choose. It's unexplainable."

But when it came to the speaking, it was Bryan, she said, who told her, "Yudith, just speak from your heart. Just tell 'em how it is."

"It's a very personal thing," she said. "Especially when it comes to the kids. Every time I watch the video we made about Manchester, I tear up, I lose it. Because I hear Delilah, one of the little girls, telling Valero to stop poisoning our community, and I feel it. Because I spend time with her, and she asks me, 'Yudith, why is our air like this, why do we have to live here?'"

"I have to keep in mind, there's a history behind why we are where we are. And I don't want to sugarcoat anything, even for kids. I think they can handle it—for them to understand, yes, there are certain systems that exist specifically to keep you in this way, so that's why you have to keep yourself educated, and that's why you have to believe that you are worth more than what people make you believe."

I told her it's important for people to hear her say those things—not only in Houston, but people in the climate movement, like those in Pittsburgh.

"I think they just want to hear their conscience out loud," she said matter-of-factly. "They want to hear me speak it."

She was right. Or, at least, since I can only speak for myself, she was right about me.

———

"My name is Grace Ann Cagle, and I was born in Fort Worth, Texas, and I lived there all my life. My parents lived there most of their lives, and my grandparents lived there most of their lives."

And how old are you?

"I'm twenty-three."

You're a mature twenty-three.

"This last year has definitely done it for me."

Grace, or Luna, as she was known in the East Texas cross timbers where she risked her life, is tall and athletic with medium-length blonde hair and an easygoing laugh. She sat in a chair across from me in the cramped space and fluorescent light of the TEJAS office on Harrisburg Boulevard in Houston. Nobody else was around—as a TEJAS "intern," Grace had a key and let us in. A large whiteboard on the wall next to us was covered with a meticulous planning chart for the Healthy Manchester Festival that afternoon. On the wall opposite, across a conference table that all but filled the narrow office, a facsimile of the Declaration of Independence hung between a large brown coffee sack from Chiapas and a stylized portrait of Emiliano Zapata.

Grace had been a biology major with a focus on ecology at the University of North Texas in Denton when she fell in with the folks from Rising Tide and Occupy Denton, and then helped launch Tar Sands Blockade in the summer of 2012. Her father is a career chemist at Alcon, a pharmaceutical company in Fort Worth, where he met her mother, also a chemist and a former middle school science teacher, who's now a "pretty successful" lawyer. "My mom's a firecracker," she said. "She parties harder than I do."

Her parents were traditional Texas Democrats, active in party politics. But Grace's grandmother, her mother's mother, was the one she really liked to talk about. "My grandmother has a PhD and taught at the big university in Fort Worth, Texas Christian, as a psychology professor—and had two kids by the time she was nineteen. She is one of the reasons I've been able to do this work, financially and emotionally supporting me throughout this—her and my grandfather. I got arrested and they framed my mug shot. My mug shot's on the fridge."

Things were different with her father. When she started taking philosophy classes in college along with biology, he was less than pleased. "It drove my father through the roof. Like, 'Asking life's big questions—and do you want fries with that?'"

I asked if she'd had a falling-out with her dad, over what she'd done during the past year. "No, it's cool," she assured me. "This has actually brought us closer. I think he's understood me better, maybe even understood himself better. He knows how dire the situation is, and he doesn't think there's anything we can do about it. But when I was in the trees, he came out to one of our protests—drove three hours out to East Texas in the driving rain. He knew I was up in the trees, and he just got in his car and drove out there. He wanted to join the protest."

I told her my own daughter's name is Grace, then nine years old. She smiled. "Yeah?"

What about religion? I asked. Did she have a religious upbringing? "Strictly atheist."

Strictly? I asked.

"Strictly."

Well, I said, that can be a kind of religious upbringing.

"Kind of, yeah," she laughed. "No, it really was." She added: "I'm a very spiritual person, actually. I practice, like, the earth."

In high school, she said, she explored Buddhism, but decided she wasn't ever going to be a Buddhist. Then she started dating the son of a Baptist preacher in Fort Worth.

Uh oh, I said, look out for those preachers' sons.

"Yeah, he was a bad one!" she said, laughing. "I almost married him. We dated for two years. He was wild."

It was a classic, Texas, Bible Belt kind of story, I said.

"I spent a lot of time with his family," she said, becoming serious again. "They were pretty progressive. It was my first experience with progressive, liberal Baptists. We'd talk about climate change, we'd talk about the death penalty. And I was like, wow, we really have a lot in common." I said it sounded like they would be on the same page with Pastor Kyle and the Austin Heights church in Nacogdoches. "They would."

So, what was it, I wanted to know, that really led her to risk her neck, sitting eighty feet up in a tree, trying to stop that pipeline?

"What am I going to do to fix climate change?" she said. "I'm not just going to sit in a tree. I mean, nobody buys that. I know it's going to take massive, almost unimaginable system change—or everything's just gonna flood." She agreed with me that the sort of system change required isn't going to happen through politics as usual. Something's going to have to force it.

"That's what originally brought me into the organizing work," she said. "That's why I was willing to sit through four-hour meetings, when we were starting Tar Sands Blockade, this really grueling work. I was like, all right, I'm going to do some climate organizing."

"But I guess what really led me to spend, like, countless nights, up all night climbing in trees—on ropes, *really* scary—was the actual *place*."

Grace said she'd come to see two converging ideas, or motivations, forming in her mind. One was the climate crisis, which led her into the organizing. The other, she said, was the experience of actually being out on David Daniel's property near Winnsboro, where the blockade began. "Just this overwhelming sense of place," she said, "that I wanted to defend."

"I mean, they've destroyed a lot of it," she said, referring to Trans-Canada. "But it's incredible. It's cross timbers, part blackland savannah, and then post oak, that mixes with the savannah. It's pretty flat land, foresty, but not piney like Nacogdoches. It's oaks."

She said you can look at an ecological map of Texas and see that there's only a small piece of it left. "It's a very endangered habitat, and there's at least two endangered species that live there—the ivory-billed woodpecker and this black and red beetle, I don't remember its name."

She tried to describe what it was like at Daniel's place, "building this tree fortress, doing all the rigging and tying. Not a single nail went into a tree. And I spent a lot of time just sitting in the trees, with the smell of the trees on me, and I saw the ivory-billed woodpecker and knew that this was its habitat. And this was probably the most beautiful place in Texas I'd ever been. There are two springs, with little creeks running through the property, and even in July it was probably only eighty-five degrees underneath those trees."

"These woods in East Texas," she said, "are sacred, literally, like a sacred place for me. I have long blonde hair, and I actually cut it off in the trees and buried it there in the earth to defend it. There was a lot of woo-woo hippie stuff that went on out there!"

I told her I didn't know much about woo-woo hippie stuff, but I knew what she meant about sacred places.

"Yeah, it's like they're putting a Keystone pipeline through Walden Pond," she said. "That's basically how it was."

She'd been up in the trees for about a week, in late September 2012, when the Keystone construction reached Daniel's property. She was the last of the tree sitters to get up on a platform. She'd noticed that the south side of the property was undefended—"I was going, holy shit! there's a massive hole in our fortress, and there's no one to sit up there!"—and so she worked for forty-eight hours rigging a platform, without sleep. "It was this triangle with parachute cord woven in between it, like a dream catcher."

"I'd been through a couple thunderstorms," she said, "and it was windy—it was a little hairy."

Sure enough, TransCanada's machines came up from the south.

"They came over the creek," Grace said. "And I was watching them. They had a feller buncher—it grabs the trees, cuts them, and throws them. And as they cross the creek, they're coming like ten feet, twenty feet away from me, practically at the base of my tree—and I thought they were going to kill me. I thought they were. And I wasn't surprised. Why would they care about me? And so I jumped onto this traverse rope, and I'm dangling there, wearing all black with a mask on my face, screaming, 'Go away! Get out of here!' They stopped their machines and all came over and looked up at me, and were like, 'Whoa!'"

"I spent like six hours dangling there, in a harness, because I could protect two trees at once. I knew that if I moved, they were there with their machines, and they wanted to cut those trees down—and they would come onto David's property and through our section. And at one point, the sheriff comes over—I'm having this conversation with the Wood County Sheriff—and he's like, 'Why don't you just come down?' And I told him, because I care too much about this place. But it wasn't just about David Daniel's home anymore, it was about the entire planet."

"That's when it all reconverged for me," she said. "At that moment, I was in total solidarity with the people in Canada, around the mines, and the people here in Manchester. It was no longer just 'climate' and 'home,' global and local. It was the whole thing."

I'd been absorbed in her story, and when she said Manchester, I was suddenly back in the TEJAS office. I wondered if that sense of solidarity had anything to do with climate justice, and what, if anything, that meant to her.

"I don't have an in-depth background on justice issues," she said. "I'm more an earth person, a holistic person, and the fact that irreparable damage is being done to the earth—that's not just."

"We are part of the earth," she said. "Ultimately, all of us will be affected. But the most vulnerable people in society—those without political, financial, class privilege, race privilege—are the people affected first."

It was a textbook answer. So I asked her why it was strategic for Tar Sands Blockade to engage with TEJAS and help organize there in Manchester. This time her words sounded less practiced, more passionate.

"We're organizing in Manchester because *it needs to be done*," she told me. "It's not a means to an end. We're working in Manchester because we want to, with all of our hearts, and feel called to—but it's also one of the hot spots where it aligns with the climate work, where our goals align with community needs. Just like at David Daniel's. I would've done that at David Daniel's anyway, just because I could. And I would work in Manchester even if climate change didn't exist."

Grace told me that the seven months she'd been living in Houston, working in some of these communities with TEJAS, canvassing, conducting health surveys, helping organize the festival, and just talking with people, had changed her perspective. She understood the science, she said. "But even as urgent as the climate crisis is, we're not going to solve it by ourselves. So whatever it's going to take to be able to work with other people, even if it takes another ten years, that's what it takes. And that can be harder than sitting in a tree, or locking yourself to something."

And what does it take to sustain that struggle, I asked her, to stay in the fight?

"I think it takes a spiritual fountain, and calling. If it weren't for the connection I feel to the earth, I would quit. Because it's really difficult. I'm not directly fighting for my house or my family, but the earth is in imminent danger, and so are my friends, and it takes this sense of being united, together, and this feeling of caring not just for each other but for the earth. And where does that come from? It comes from somewhere deeper."

"The whole reason I'm doing this," she said, "is the continuation of life. I wish there was no fossil-fuel industry, and I could just live on a farm and have a bunch of babies, 'cause honestly, that's all I want to do. If everything wasn't going to hell, then that's what I'd like to do."

"You can't just give up," she said.

Please don't give up.

"I can't. I can't give up."

———

Before I left Nacogdoches that July, the blockaders gave me directions to their camp outside of town, just north a few miles on Highway 259 ("there's a taco stand on the right and a radio tower . . . look for the white mailbox . . ."). I arrived midmorning, and the four I'd interviewed at the church were the only ones there. Murtaza showed me around. There was the tiny, ramshackle, flea-infested house that was used as a make-shift HQ, and the communal outdoor kitchen under a blue plastic tarp, which had served fifty or more at a time. There was the outhouse that one of the Austin Heights members had built for them. As we walked a footpath into the woods in back, I saw the few remaining tents, and an ingeniously rigged, if less than private, "shower." Some climbing tackle still hung from a large tree. Nearby was a big pile of buckets and containers once used for hauling water.

Murtaza thought the camp now had a vaguely postapocalyptic look— or maybe, I thought, like a guerrilla encampment after the fighting had shifted to new ground.

After my tour, I sat down with Fitz at the picnic table by the kitchen, next to a campfire he was tending. I asked him if the "blockade," as such, was over with.

"It's hard to define over," he said. "When I got here, blockading was as direct action as direct action can get. That part of TSB, in Texas, I think is done."

What about building a deeper resistance that can go forward?

"TSB didn't come here to create a resistance," he said. "That resistance already existed. We partnered in that resistance, and we're still partnering in it."

Had TSB strengthened that resistance?

"Without a doubt. As far as resistance is concerned, we are *far* from done."

As it turned out, there in Nacogdoches, the resistance would have to come less from TSB, as it scattered to the four directions, and more from the people of East Texas.

Thunder and a drenching rain had rolled through Nacogdoches and the air was rainforest damp. As I drove south past the university on North Street, in the center of town, astonishing back-lit clouds broke up in the west and the late-afternoon sun shot Jesus rays onto the steaming land-scape. I'd been in Texas more than a week, and I'd just finished another long day of interviews. My head was heavy, completely saturated, and I could swear my body felt the weight of all the lives I'd pried in upon—the good people with their big-hearted laughs and sad eyes, the weariness and wariness in their voices, which they couldn't hide.

I needed to get out of the city limits, clear my mind. I turned west onto Main Street, where the old proud sign declares Nacogdoches the "First Town in Texas" and the county courthouse occupies the southwest corner, an ostentatious modern complex with a plaza out in front. The street dipped and crossed Bonita Creek and the railroad tracks, then passed through the hard-scrabble neighborhood on the west side. At the edge of town it turned into the Douglass road, Highway 21, and I kept on going the fourteen miles, into the sun, through the lush rolling woods and pastures to Douglass, Texas, population 1,100.

It was time to see the pipeline. Or if not the pipe itself, now hidden in the dirt, then at least the clear-cut scar that snaked the length of East Texas—from Lamar County, where my dad was born, to the Gulf of Mexico. My first night in Nacogdoches, Kathy and Steve DaSilva, core members of the NacSTOP group, had invited me out for dinner at their house in the country east of town, and they'd told me about Douglass, where Keystone passes perhaps a hundred yards from the local school.

Sure enough, when I turned into the school and followed the narrow blacktop lane down past the ball field, there it was. I pulled over at the wide, long clearing in the woods, the dark soil not yet overgrown by the weeds. But I stayed in the car, keeping an eye in the mirrors. (Fear of surveillance, or God knows what, is contagious, I'd learned.)

The sight of it was anticlimactic—dispiriting. A mood of futility seemed to seep up from the ground. It taunted me. The silence was oppressive. A single white sign stood sentinel, facing the road:

TransCanada
Keystone Pipeline
WARNING
ADVERTENCIA
High pressure oil pipeline
Oleoducto de alta pression

———

"You know what it feels like?" said NacSTOP cofounder Vicki Baggett, back in town, referring to the way the national climate movement had never really invested in the fight to stop Keystone's southern leg. "It's the same reason Obama blessed that pipeline in Cushing to come south. We're rural, we're poor, we're not very educated, we're not very connected. If you look at a map, every pipeline in the entire world goes through East Texas. I think they feel like it's just—a lost cause. That's what it feels like." She noted another pipeline, the Seaway, that was coming on line to carry Canadian tar-sands crude to the Texas Gulf Coast. "It's gonna be a superhighway of tar sands through East Texas."

We were eating lunch at Java Jack's, the all-purpose and not-too-funky café with free Wi-Fi that served as TSB's nerve center when they were in Nacogdoches. Vicki and her husband, longtime Nacogdoches residents, have two grown sons, and she said the blockaders who spent time at their house became like a whole new set of kids. "I got to cook for them, I got to be Mom. They have a real dear spot in my heart." But there was also fear. Her husband, who owned and operated a tree service, was afraid there would be retaliation against the business, or worse. "This is how they win," Vicki told me. "It can get dark." Intimidation, fear of being watched and followed, tension within marriages—all of it just wears people down. That's why a community like the Austin Heights church was so important.

Vicki was active in the local Sierra Club chapter when she learned about Keystone XL in 2010, and together with a few others decided to start NacSTOP. She said they had their meetings, wrote their letters, built their e-mail list, signed their petitions. And it probably never would've gone beyond that if the blockaders hadn't arrived. "This has been such an intense time," she said. "And whatever comes next"—she paused. "They really grabbed people's hearts," she said. "They changed people. And we're not going to change back, because we're aware now of things that we never were."

"They've had a real lasting impact, on me personally," Vicki said. "My interaction with them has helped me grow spiritually. And I've heard other people at Austin Heights say the blockade has helped them grow spiritually. You know, 'maybe those kids are onto something.' I think they've stretched us, and I think we've grown."

Almost a year later, I talked with Vicki again, and she told me that NacSTOP was rethinking its purpose, expanding its mission—and it soon changed its name to Resilient Nacogdoches. "It's about creating community," she said, "and it's no longer focused only on tar sands. We're looking at fracking now, too. We're opening up to anything that makes our community healthier and safer, and more of a community." Fracking, of course, had exploded across the state in recent years, and especially in East Texas.

Starting with an effort on emergency response preparedness, in part to educate public officials on the danger of pipeline ruptures and spills, the group had managed to place one of their founding members, Kerry Lemon, on the local emergency planning commission. Kerry's daughter, Maya, who'd also been active with TSB, told me how the Lemon family's property outside of town has fracking wells on it, along with compressor stations, chemical storage tanks, and a pipeline running down the driveway. "Because my family doesn't own our mineral rights, Exxon has been able to do pretty much whatever they want on our land," Maya told me.

(When Maya was fourteen, her dad was diagnosed with leukemia—a result, they believe, of exposure to known carcinogens linked to petroleum production.)

"It's not over yet," Vicki said. "This really is part of a very long conversation. We're creating a community, and creating a movement, and it's growing. People are learning. People who didn't used to ask questions are asking questions."

———

My last weekend in Nacogdoches, I sat down again with Kyle Childress, this time in his comfortable, book-lined study at the church, where a portrait of Martin Luther King hangs on the wall. I told him, only half joking, that it was a little intimidating to sit there under the gaze of Dr. King.

"There's a line in the book of James," Childress said, "that says the prayers of a faithful person *availeth much*. One person, one small community, acting in faithfulness, can bring healing, hope, change."

"Fifty years ago in Birmingham, Alabama," he went on, "you had all these African American kids—teenagers, twentysomethings—going up against Bull Connor's police dogs and fire hoses." He paused. "I want to be pastor of the kind of church that produces young people like that."

"Some of these blockaders," he said, "were risking their lives up there in a tree trying to block that pipeline, and TransCanada has billions of dollars and says, 'We'll just go around you. You slowed us down for a day.' Well, if that's all there is, by sheer mathematics they win."

"But I think the prayers of a faithful person availeth much," he said.

There was a sound in his voice, something in the cadences and the accent, that I knew from somewhere, a long time ago, some deep chord I'd forgotten how to hear—or maybe it had been taught out of me.

"And those blockaders," he said, "are acting in fidelity to the goodness, the rightness, of God's Earth. That keeps me going. That's my hope. And if I didn't have hope, well, I'd probably just cash it in and go do something else for a living. I mean, you know, I'm not going to be pastor of a church without any hope."

SIX

Too Late for What?

We are now faced with the fact that tomorrow is today.
—MARTIN LUTHER KING JR., 1967

Mount Auburn Street, a block south of Harvard Yard in Cambridge, Massachusetts, with its brick sidewalks, its exclusive old-boy college clubs, and its late Victorian "Gold Coast" dorms, is about as far as you can get from the piney woods of East Texas.

Unless, that is, you find yourself sitting on an ancient cast-off sofa in the small, dimly lit "library" of a former fraternity house, sunlight and street sounds filtering through lowered blinds, where a group of young people plot nonviolent direct action at the suburban offices of TransCanada Corporation—in solidarity with the Tar Sands Blockade fighting construction of the southern leg of the Keystone XL pipeline. Then you're practically in Nacogdoches.

That room is where I found myself on an afternoon in December 2012. I was there not as a journalist but to help the organizers, mostly students and recent grads from nearby schools, with media outreach and communications. It was the first time I'd been on the inside of a well-planned civil-disobedience action—the first time I'd felt the tingle of adrenaline, and the faint undercurrent of anxiety, that comes from participating in an act of principled resistance where power would be confronted, laws would be broken, and people would go to jail.

Nervous laughter punctuated the bewildering logistical checklist. This would be no ordinary sit-in. I listened with fascination as kids twenty-five years younger than me, utterly in control, spoke of "jewelry," or hardware—in the present instance, the hardened-steel chains and locks they would use to secure themselves to each other once inside the corporate office. I'd gotten to know several of the students and twentysomethings over the past summer and fall, working alongside them in the Boston-area climate movement. But in that room, that afternoon, they appeared suddenly older, mature beyond their years, and uncommonly brave.

All of my young friends were acutely aware of their privilege—and yet all of them were fighting for what they called climate justice. And not only for poor and vulnerable people in faraway places but, more and more, for themselves—and for one another. They felt themselves—people of their generation and younger—threatened and oppressed by powerful, corrupt forces far beyond their control. They grasped the urgency and scale of the climate catastrophe—and understood the role of the fossil-fuel industry and its political enablers not only in obstructing any serious efforts to deal with the crisis, but in actually accelerating us toward the climate cliff. They quoted the alarming reports from the International Energy Agency—stating that continued global investments in fossil-fuel infrastructure past 2017 would "lock in" catastrophic warming—and from the World Bank, which had recently declared that humanity is on course for warming of four degrees Celsius this century, likely beyond our civilization's ability to adapt. They saw the catastrophic trajectory to which their elders have condemned them, and they felt something like desperation—forced onto a radical path by the political and moral failures of older generations. And their analysis was, and is, painfully accurate: at this late hour, to be serious about our climate reality is to be radical. I wondered then, and often still wonder, if young people like these—and those I've met in Texas—are the only people in this country with the guts and maturity to accept what that means.

And so it was that on the morning of January 7, 2013, eight of my young friends, supported by perhaps a dozen more organizers of varying ages and experience—and with an increasingly galvanized and resolute national movement behind them—walked into the TransCanada suite on the second floor of a nondescript office park among the strip malls along Route 9 in Westborough, Massachusetts. "This is a peaceful protest," they announced to the receptionist, and then sat down on the floor facing outward in a tight circle, locked themselves together with those heavy-duty chains around their waists and ankles, and joined hands—bound with superglue. (That final youthful flourish ensured headlines like Boston .com's: "Protesters Glue Themselves Together at Westborough Office of the Company Building Keystone Pipeline.") Their names were Emily Edgerly, a twenty-year-old sophomore at Tufts; Devyn Powell, a twenty-year-old Tufts junior; Lisa Purdy, a twenty-year-old junior at Brandeis; Shea Riester, a twenty-two-year-old Brandeis graduate; Ben Thompson, a twenty-two-year-old grad student in mathematics at Boston University; Ben Trolio, a twenty-one-year-old senior at the University of New Hampshire; Alli Welton, a twenty-year-old sophomore at Harvard; and Dorian Williams, a twenty-one-year-old senior at Brandeis. On the carpet, just inside the reception lobby's glass doors, they spread a banner that read: STOP THE KXL PIPELINE NOW. In the photo taken by their support team and distributed to the press, they're sitting in that circle on the floor dressed as though for job interviews. Eight polite young people concerned for their future.

"We stand together as representatives of a desperate generation," they wrote in a statement posted on their website (january7th.wordpress. com). "Today, we hope to present our political leaders with an example of the courage needed to confront the climate crisis by putting our bodies in the way of corporations whose activities threaten our society." They acted in solidarity not only with the Tar Sands Blockade in Texas but also the First Nations peoples in Canada resisting the extraction of the Alberta tar sands and the poisoning of their sacred ancestral waters and land. And, the students wrote, because the tar sands represent "even greater destruction" to come from "unabated climate change," they acted "in solidarity with all humanity."

Their principles didn't seem to impress the Westborough Police, who tried and failed to cut through the chains—claiming they damaged the equipment they'd brought in for the job—and shuffled them chain-gang style off to jail, where they spent several hours before being charged with trespassing and released on bail. In the end, with pro bono representation by the National Lawyers Guild, they paid $3,439 in restitution to the town of Westborough, and served six months of probation.

Two months later, more than a hundred of their friends and supporters (including me) went back to Westborough, where twenty-five people—carrying a mock coffin and singing "They're Digging Us a Hole," a foot-stomping dirge written for the occasion—sat down at the entrance to the TransCanada suite, and were arrested.

More important than any particular action and its outcome is to understand what led these young people to it—not their media talking points, but how they understood, as individuals, what they were doing and why. And not just what they were thinking, but how they felt. Their fear, anger, love—their willingness to sacrifice and to lead while their elders slept. And so a few weeks later I spoke at length with three of them, all deeply involved in the New England climate-justice group Students for a Just and Stable Future (SJSF) and the grassroots network 350 Massachusetts.

Harvard sophomore Alli Welton was studying government and the history of science and was a leader of the Divest Harvard campaign—though she would soon leave Harvard to focus on climate and social-justice organizing, waiting tables to pay her rent. She grew up in a small town in eastern Washington, where her father was a physician. Arriving at Harvard in 2011, she dove in with SJSF and spent time at Occupy Boston, getting to know a number of older climate activists, some of whom had been arrested at the White House protesting Keystone that August. "The arrest record seemed like an exclusive symbol of commitment," she told me, and for a while she had "a distorted idea of civil disobedience." But after taking courses on civil rights history and the media, she gained a better understanding, she said, of its "strategic value."

"These companies are waging war on us and our lives," Alli said. "And we have to fight back—somehow. It's surreal sometimes to realize that this is the world we live in, that we have these giant corporate tyrants who are controlling our lives and sacrificing us for their profits. It's insane that we live in this society."

"One of the seniors involved with SJSF said something like, 'You know, I think I could die of climate change; that could be the way I go.' And that stuck with me," Alli said. "It's very possible that we could be the generation watching our society crumble away. Sometimes I walk around Harvard late at night, with all these huge, fancy buildings, and think about what Rome was before it fell."

Alli had long thought of climate activism "primarily as solidarity," she told me, "and helping reduce inequality in the world, which is something I've cared about ever since I was a kid, growing up privileged in a really poor town." But recently, she said, it had also become about "self-preservation."

"You learn about marginalized groups in society," Alli said, "and how their voices don't have as much power. And then suddenly, you're like, *wait, that's exactly what I am with climate change.* I'm like the helpless kid here just begging the older generations to save me. And what the hell is that? That's a hard—I don't like begging."

Tall, mild-mannered Ben Thompson was in his first year of graduate studies in mathematics at BU. He grew up in New Hampshire, his father an accountant and his mother an elementary school guidance counselor. During his freshman year at Cornell College, in Iowa, deeply affected by reports of the devastating 2009 wildfires in Australia and their link to climate change, he began thinking seriously about nonviolent direct action. "I remember looking up the Iowa legal code trying to figure out what would happen if I sat down in front of a coal plant entrance," he told me. "But I was in Iowa with no support, and it was a big scary step, so I never ended up doing it."

"I remember reading a quote from a climate scientist, something to the effect of, 'It's not quite time to chain ourselves to the statehouse, but we are getting close'—and I remember thinking, 'It sounds like it's time to me.' And I think that decades from now, people are going to say, 'Of course you were going to chain yourself together inside a TransCanada office. Why wouldn't you?'"

I asked him whether he'd been scared.

"The fear that I felt around the action pales in comparison to the fear I feel around climate change," Ben told me. "I've spent sleepless nights and had panic attacks at 4 A.M., thinking about—you know, reading reports, and just thinking, like, are we really doing this? Am I really expected to read this and then go do my studies—like nothing's happening? This is insane."

"I think our only asset as a movement is that we have everything on the line," he said. "That's our only asset. I'm willing to risk my life for this."

Three weeks after the Westborough action, I stood yards away as Ben spoke to a crowd of 1,500 at a boisterous protest in Portland, Maine, against a proposed tar-sands export pipeline from Montreal. In attendance were Portland's mayor and its representative in the US Congress. "I'm ashamed to say that it took me nearly eight years to tell my dad that I am a climate activist," Ben said with great passion, his voice cracking. "I finally had to tell him, because I had to say that I was going to wrap hardened steel chain around my waist until the police cut it off."

Dorian Williams, who was then majoring in anthropology at Brandeis, grew up in Chicago, where both of her parents are college professors. With her warm smile and youthful features, you might not take her for a hardcore activist experienced in the ways of nonviolent resistance. But that day in Westborough, at the age of twenty-one, she went to jail for the fourth time.

Her first arrest was in April 2011 in Washington, DC, at the Department of the Interior, protesting mountaintop-removal coal mining and other extreme fossil-fuel extraction. She had just attended the three-day Power Shift conference, which she described to me as her galvanizing moment. In particular, she recalled a succession of three keynote speakers: Lisa Jackson, the EPA administrator; Bill McKibben of 350.org; and Tim DeChristopher, who had been convicted the month before in federal court for disrupting the Bureau of Land Management auction in 2008, and was awaiting his sentence.

"Lisa Jackson said, 'Look, I'm doing everything I can, but it's not going to happen from the inside. People need to push from the outside,'"

Dorian recalled. "And then that was exactly what Bill McKibben went into. But he said, 'Look, we've been doing the easy stuff—we've been doing these big days of action—and it hasn't been working.' And then Tim DeChristopher got up and said, 'Look, I went out and did this action, this is what we need to be doing.' He said Power Shift can't happen like this again—we can't keep coming here and not doing anything. 'There are ten thousand people in this room,' he said. 'We need to all go down to West Virginia. If we sent enough people down there we could end mountaintop removal.' And I remember standing on my chair, making a promise to myself that if what we were doing wasn't working, then I needed to be doing something different."

On the last day of the conference, thousands poured onto the street in an "unauthorized" march to the Department of the Interior. "Eighty people rushed the doors and filled the lobby," Dorian remembered, and she was one of them. "And I was in this huge room surrounded by all this energy, and thinking, 'What am I going to do right now?'" Sitting down with the others, she asked herself, "Am I going to get up and leave, or am I going to stay?" She stayed.

"It was a huge identity shift for me" Dorian said. "I was committed in a different way. Like, if it's not me, who is going to do this? The people in West Virginia can't afford to travel to DC and get arrested. A lot of people can't ever afford to get arrested. You know, I'm white, I come from a privileged background, if people like me aren't willing to take this risk, then who on earth can afford to take these risks?" Her next arrest was in front of the White House in August 2011, along with the 1,252 others protesting the Keystone pipeline.

Dorian's third arrest came in July 2012, on a mountain in West Virginia, where she and others—organized by Radical Action for Mountain People's Survival (RAMPS), a nonviolent direct action campaign against Appalachian strip mining—locked themselves to a truck at the Hobet mine in Lincoln County, the largest coal mine in the state. She spent ten days in jail, unable to make the $25,000 property-only bail, until sentenced for trespassing and fined $500.

"Walking onto a mine site and locking to a piece of machinery was unlike anything I'd ever done before. I mean, it was a level of personal risk. It wasn't in DC. It was actually where things can go wrong. You don't know how the cops are going to respond, how the miners are going to respond. You're confronting people who have every right to be pissed off at you."

"I was treated fine," she said. "Some of the other people I was arrested with were not treated fine." Watching the police take two of the male protesters into the holding facility was, she said, "the most violence that I've ever personally witnessed." They were being noncompliant, she said, and the police "were just really not careful with their heads." She saw the protesters yanked out of a van, "so the guy hit his head on the metal van, and then again on the ground. And then a bunch of them were trying to pull him through the door of the processing center, and he hit his head really loudly on the door frame. I thought they'd knocked him unconscious."

"Leading up to the action was even scarier," Dorian said. "There were a couple points that were really new degrees of fear that I had not felt before. Because I had no idea what was going to happen. That was the scariest thing. I didn't know what was going to happen to me. I mean, the night before was terrible. I was desperately trying to find reception to contact my parents and my support people from home, and trying to figure out whether to go through with it. I was less worried about my personal safety at that point, and more, just—I didn't want to be a burden on my family."

"And eventually it came down to the same thing as at the DOI: if someone who has as much privilege as I do can't take these risks, then who's going to stand up for this stuff?"

"But I want to be careful not to glorify those who get arrested," Dorian told me. "Every time I've been arrested there's been twice as many people behind me, and those roles are so important. I want to make sure that everyone out there who can't get arrested—because not everyone has that ability—still feels like they have every bit as much influence and ability to be powerful in this movement."

Dorian told me that she had begun to see her motivations in a new way. "When I started out in this movement, it was because of the injustices I was seeing. Going to West Virginia, that was in solidarity with those people who—our government is basically allowing a war to take place on them. These companies are blowing up their homes. If another country had done that, that would be an act of war."

Those injustices, to be sure, were still part of her motivation, she explained. What was new, she said, was a "deeper acknowledgment of what's at stake, and what's on the shoulders of my generation." Climate change, she said, "has the potential of being the single most trying thing for our species in thousands of years, and it's falling on our generation to transition to whatever that's going to look like, that great unknown, in a matter of our lifetimes and our children's lifetimes."

"It's gotten a lot more personal for me," Dorian said. "The fear I've experienced in my activism has never been greater than the fear that I have of climate change. It's visceral. It's very real—to a degree that, on good days, it makes me fight, it gives me a fighter's instinct, which is not something that I would naturally have.

"But on bad days, it makes you want to run away. I know people who want to build isolated farms in the middle of nowhere, to try and escape it all. And sometimes it's paralyzing. I've gone through literally bouts of— not depression, but very serious, very low states of being, for hours to days to weeks at a time. It comes and goes, and you have to wade through it, because it's never going to go away. You just have to work through it, so that you can keep fighting."

I was curious if Dorian had experienced her struggle on what she'd call a religious or spiritual level.

"My mom comes from a Protestant background, my dad comes from a Jewish background," she told me, "and I kind of grew up with a lot of religion and no religion at all. But I think, to me, it's less about religion, and more about community—and empathy."

"In some ways," she said, "the beauty of this whole thing is that because it's the greatest challenge, it's also the greatest opportunity to come together and reclaim community, and a global community, and allow people to think beyond themselves. My generation hasn't had enough of that. I think that we're really atomized and separated and segregated from ourselves, and there's a hole, and an absence. I've not felt so connected to the people around me as I have in this movement."

On a warm Sunday evening in September 2013, Tim DeChristopher appeared through the door at the back of First Parish Cambridge, the historic church across the street from Harvard Yard and the offices of Harvard's president. (A banner above the church entrance read "We Divested from Fossil Fuels. Your Turn, Harvard.") It's the same church in which Ralph Waldo Emerson delivered the Phi Beta Kappa address known as "The American Scholar" in 1837, the year of Henry Thoreau's commencement. I was sitting in the front row of the brightly lit and nearly empty sanctuary—with its high rafters and the pious New England minimalism of white-painted pews and dark wood—tapping on my phone. Someone fiddled with a laptop projector a few feet away, last-minute preparations for a community event to raise money for the "Westborough Eight," as my young friends were now known. Tim was the night's featured speaker.

I was the first person Tim saw as he walked through the door to one side of the pulpit. We'd never met, only e-mailed and tweeted, and when he saw me, he smiled. "Hey, Wen." We shook hands—"Hey, Tim, it's great to see you, thanks for being here"—and we sat down on the front pew. The event was supposed to start in a few minutes, and people were trickling in. "Where is everybody?" he said.

There he was, Bidder 70, my "new abolitionist" in the flesh, a bona fide movement rock star—in his first weeks at Harvard Divinity School, five months out of federal custody, minutes before his first climate movement event in Cambridge, a fund-raiser for students inspired in part by his example to engage in nonviolent direct action—and the place was practically deserted. Where was everybody, indeed? In Salt Lake City, he whispered, the church would've been packed.

Welcome to Cambridge, I thought.

To be fair, First Parish is a large room; it can seat some 600 people, and no one expected it to fill up. By the time things got started and Tim rose to speak, there were probably 150 or 200 people in the audience. A lot of them were familiar faces, young and old, friends from 350Mass and

Better Future Project and SJSF, but many were new. Not a bad showing, actually, for a Sunday night in Harvard Square.

Tim got a standing ovation, and he launched into his talk, not from the pulpit but at a lectern on the floor at the front, wearing jeans and an open-collared shirt and blazer, clean shaven to the top of his scalp. He spoke easily yet earnestly about commitment—the kind of commitment it takes to do what the Westborough Eight had done, and the commitment we need from a movement that will be there for people who take those risks. So that when an activist puts her body and her freedom on the line, she knows there's a community that has her back. He talked about the strong community in Salt Lake City, especially around the Unitarian Universalist church where he was a member, and the climate-justice group Peaceful Uprising he cofounded, and how they supported him before and after his action, and throughout his trial, and while he was in prison.

He talked about the importance of civil disobedience, and how essential it is now to go beyond the choreographed "photo-op" actions we'd seen at the White House and elsewhere—and how important it can be for activists to go all the way to trial, and to go to jail, even prison, for their principles. How he'd learned that a trial can be a powerful movement-building opportunity.

The audience listened intently, and when it came time for questions, I stood up. I wasn't sure if Tim was aware, I said, but it looked as though we were going to have a climate trial of our own, right here in New England. Last May, I said, Ken Ward and Jay O'Hara—friends of many of us in the room—had anchored a small wooden lobster boat, the *Henry David T.*, in the path of a coal freighter at the Brayton Point power plant, blockading forty thousand tons of West Virginia coal for a day. Now they were getting ready to go to trial and faced what could be serious jail time. I asked Tim if he'd commit to supporting them. Of course he said yes. I'd piqued his interest. And then I asked him another, more personal question.

I told Tim that Ken Ward had a thirteen-year-old son, the same age as mine, and I wondered how Tim would explain to a child that his father was going to prison because of an action like Ken's and Jay's. I watched Tim's brow furrow. He said he didn't know, exactly, what he'd say—but offered that there were probably others in the room that night, especially parents, who could no doubt answer the question better than he could, as he didn't have children. And then he added: This is what a movement does. It forms communities that can hold people at times like these—and that can help explain to a kid, this is why your dad has to go away for a while, and we're going to be here for you.

The Westborough Eight had a successful fund-raiser. They were even able to send $1,612 to the "MI-CATS Three"—Barbara Carter, Vicci Hamlin, and Lisa Leggio—who faced felony charges and up to three years in jail for their direct action to stop an extension of the Enbridge tar-sands pipeline that spilled into the Kalamazoo River in 2010.

Soon after that night, I connected Tim with Jay and Ken.

"I'm both calmer, now, and more radical," Tim said to me. "I mean, personally I'm more at peace now. And having gone to prison, I'm more politically radical. After spending a couple years in the custody of the government, I have a better picture of the nature of the government we have, and I'd say that it's my goal to overthrow our current form of government."

Tim sat across from me at a small table in the aptly named Shay's (think rebellion), an English-style pub and one of my old haunts, across the street from the Harvard Kennedy School—where one can earn a graduate degree in the administration of government. But Tim, of course, was a student at the Divinity School, located appropriately enough in the far opposite corner of Harvard's campus. It was an afternoon in mid-December, three months after our first meeting, and the first in a series of long, searching conversations during the better part of a year.

I asked Tim if he's an anarchist. "No," he said. And unlike the so-called green anarchists, he noted, "I don't think that civilization has been a mistake. I think there are a lot of wonderful things about our civilization that are worth keeping."

Overthrow is a pretty strong word, I suggested.

"I mean our current system of corporate rule." We need to force a new constitutional convention, he said, "to form a new government in this country."

"Ending corporate personhood," Tim said, "and declaring that money is not speech—that in itself would create revolutionary changes in the way that our government and our whole society works." And those things, in his view, will probably require a new constitution.

I raised my eyebrows. Cleared my throat. He wasn't joking.

"I also recognize that in a constitutional convention, everything is on the table," he said. "And there are deep, fundamental divides in this country—there are genuinely different interests, genuinely different worldviews, at a fundamental level—and I'm not sure a convention would resolve itself."

What would it take even to get us to that point, I wondered.

"I don't necessarily know what it would look like, or where things would go," he said. "I say 'overthrow' because I recognize that the people currently in power are not going to willingly transfer power into a democratic form. In recent history, nonviolent revolutions have been far more successful than violent ones—but it still takes that kind of pressure."

Did he really think that sort of change could happen without bloodshed—in a country armed to the teeth?

"Is it conceivable that we'd get to a constitutional convention without progressives forming a militia? Absolutely."

I laughed.

"Is it conceivable," he went on, "that we'd get there without some activists getting killed, either by the state or by corporate hit men, or whoever it may be? Not really. I do think activists will lose their lives, as they have in almost every major struggle for justice."

No laughter. That thought, I said, treading more carefully, must have crossed his mind before—on a personal level.

"Yeah," he said. "Yeah."

When did that start? I asked.

"When people started comparing me to Martin Luther King."

I winced.

"Especially once I turned thirty, and people started referencing Dr. King in relation to my story, I was very conscious of the fact that Dr. King never lived to see the age of forty."

How old are you now? I asked.

"Thirty-two," he said. His birthday was November 18.

I steered the conversation back to where we were sitting. I wanted to know why he was there at Harvard—and why he'd chosen divinity school, of all things. I knew he'd felt close to his congregation in Salt Lake City. Did he want to be a Unitarian Universalist minister, a pastor of a church?

"That's the plan," he said.

A clerical collar, or any other religious vestment, isn't necessarily among the images most people call to mind when they think about climate change or the climate movement—much less the radical edge of the climate-justice movement. But then, maybe it should be. (And if Pope Francis has any say in the matter, it very well might. His unprecedented, and long-awaited, papal encyclical on climate change challenges the assumptions of religious conservatives and secular progressives alike.)

I remembered reading two years earlier, in Tim's long interview with Terry Tempest Williams, that he'd been through a dark time—what some would call a dark night of the soul—when he first came to grips with what climate science is saying and what it means for people of his and younger generations. I told Tim that it seems impossible to have an honest conversation about the climate movement without acknowledging how late the hour is, without confronting that reality. I wondered how much he thought about it, and whether it had anything to do with why he was there in Cambridge.

Tim looked me in the eye. "Why I'm here is very closely connected with the fact that it's too late to prevent the collapse of our civilization—that we are already committed to a path of chaotic and rapid change. And I'd say"—he paused, exhaled. "That impacts all of my thinking."

"Accepting that reality does put all of us into that dark night of the soul," he said. "It pushes us into a period of despair. And climbing out of despair—I think people do it in a lot of different ways. Most of the people

I know who are most committed to working on climate change have been through that. But as diverse as that experience is, I think it's always a spiritual experience. Climbing out of despair is always a spiritual process."

I asked him what he meant by spiritual.

"I mean grappling with issues much larger than yourself. Grappling with existential questions, about the meaning of our lives and our existence. That's got to be a part of overcoming that despair."

"For me, that process of despair looked a lot like mourning," he said. "It was like I was mourning for my own future, and mourning for the future of everyone that I care about. And I think that, as well, is a spiritual process, and going through that tends to connect people to the spiritual nature of this larger struggle—of where we're all headed, and what we have to do as a movement."

"Our job as a movement," he went on, "is no longer just about reducing emissions—we still have to do that, but we also have this new challenge of maintaining our humanity as we navigate this period of rapid and intense change. And with that challenge, with that job, we can't avoid the spiritual aspect of what we're doing. We can't avoid talking about our most fundamental principles, and our most fundamental values, and the things that we want to hold onto the most. We can't avoid talking about our larger worldview, and our vision for the world—what we want to create, what we want to build in the ashes of this world."

"If you're connecting to real values," he said, "that has to come from some kind of spiritual place—you're connecting to something larger than yourself."

Grappling with those kinds of questions, I suggested, means confronting our situation on a level that's more than just intellectual—and that can be a tough sell these days, especially at places like Harvard.

"We're facing this challenge," he said, "at a time when the left has been solely intellectual for the past generation, has embraced a very short-term pragmatism, and shied away from articulating values, a vision, or a worldview."

"Look at the debate in 2009 over the Waxman-Markey climate bill," Tim said. "Nancy Pelosi stood on the floor of the House and said this bill is about four things: jobs, jobs, jobs, and jobs. And then a year later she wondered why young people wouldn't come out and vote for Democrats, when their survival wasn't even in her top four."

Maybe it's understandable, I said, given the magnitude of what we're facing, that the climate movement has resorted to telling ourselves and others certain useful fictions—that we can "solve the climate crisis," or "preserve a livable planet" without deep, radical change.

"There are very few things that make me more hopeless," Tim said, "than a movement based on useful fictions. What's the point of a social movement that can't tell the truth?"

Are the fictions useful, though? I asked.

"No," he said. "The only way in which they're useful is to help people cling to false hope. But that false hope is just holding people back from the effectiveness that comes after they've gone through despair. I don't think you can be effective at fighting the real threats that we face if you refuse to deal with the real world."

I asked Tim what a movement that has given up such false hope would sound like—what would it say? Instead of building a movement to "solve the climate crisis," what are we building a movement to do?

"We are building a movement for climate justice," Tim said. "That's still a relevant concept, a relevant goal—to defend the right of all people, and not only people of all races or nationalities but people of all generations, to live healthy lives and have both the agency and the environment necessary to create the lives they want. We are building a movement to hold onto the things about our civilization that are worth keeping. We are building a movement to navigate that period of intense change in a way that maintains our humanity."

"But people aren't gonna do very well, going through that pit of despair—and coming out on the other side at a point of powerful and effective action—on the other end of an Internet connection," Tim said. "It's gonna have to be a movement that holds people, that has relationships, that deals with people as people. It can't be a movement that deals with people as numbers on a petition, or that treats people as resources."

A big part of what the movement needs to do, Tim said, "is get people to really connect with their values, what their vision is, and what they're willing to do to create it." And that has to be conscious. In a sense, he said, "there's a kind of spiritual practice that gets people ready for those moments when they have an opportunity to exercise their power as citizens. People have to be ready for that."

When asked why he was going to divinity school, Tim said, he often told the story of the jury selection for his federal trial in Utah. He watched as the judge called each juror into his chambers, and said, "Your job is not to decide what's right and wrong here. Your job is to listen to what I say the law is, and to enforce it, even if you think it's morally wrong. Can you do what I ask you to do, even if you think it's morally wrong?"

"And if they didn't say yes," Tim told me, "they weren't on the jury. I watched one person after another say, 'Yes, your honor, I'll do whatever you tell me to do, even if I think it's morally wrong.'"

"I thought about that, in the context of where we're headed," Tim said. "You know, all the struggles and the disasters that we'll face, all the things that those in power will do to cling to their power, and all the things that we as citizens will be asked to do, by our government, against one another. I saw that dark potential there, and I realized the importance of connecting people to their own moral agency. Without that faith in our own moral authority, any tyranny is possible. And looking at our future, that's terrifying."

"That became a driving part of what got me here," Tim said. "We need to be morally ready."

That's a fundamentally different challenge—a deeper, more radical challenge—than "solving the climate crisis," I said.

"That's why I'm here," Tim said. "You asked what an honest movement would sound like. I think it would sound spiritual." He smiled. "I think it would sound a bit like a church."

———

Tim DeChristopher was born in the West Virginia coal country, in a little town called Lost Creek, before his parents moved down the road to West Milford. He has one sibling, a sister, a couple years older than him. His mother was active early in the fight against mountaintop-removal strip mining, but when he was eight, the family moved to Pittsburgh. His father worked his entire career in the natural gas industry, climbing from engineer to the executive suite, and when his company was bought by Dominion Resources, he was offered a comfortable early retirement package while Tim was still a teenager. Tim's mom went back to school for an MBA and worked in finance for Ford.

Tim went to high school at Shady Side Academy, near Pittsburgh. His senior year was a difficult one, and pivotal. A serious athlete, he played football (offensive guard and defensive tackle) and wrestled, but he developed a painful shoulder injury in his junior year, and in the fall of his senior year he found out his shoulder was broken and he could no longer play sports. The day after the doctor told him the news, Tim found himself sitting in the senior lounge after school, unable to go to practice, wondering what to do with himself.

"As I'm sitting there," Tim told me, "one of my teachers, a young guy, sat down next to me. He's someone I had a pretty good relationship with, and we started talking about my identity and stuff, and the conversation ends up going in the direction of meaning-of-life questions. And he was very much a Christian."

"I grew up without any religion at all," Tim says. "My family never went to church." But at the age of eighteen, he and a group of other nonreligious seniors started getting together with this teacher to read and study the Bible in a serious way. "And towards the end of that year," Tim said, "I ended up accepting Christianity."

Tim went on to Arizona State University, in Tempe, where he got involved in campus Christian groups and started going to a start-up, nondenominational, evangelical Christian church. "One of the hip, Christian-rock types," Tim said. "Young people. It met in the evening in the basement of a nightclub."

"It was very evangelical," he said, "but it wasn't right-wing evangelical. It wasn't politically conservative, but it was religiously conservative. And that became a fairly big part of my identity, as a serious Christian. I was baptized, when I was nineteen, my first year at Arizona State. All the way underwater, full immersion." (Tim and I have this in common, though I was twelve when I went under.)

At ASU, Tim also started identifying as an environmentalist, starting an outdoor conservation and recreation club, and spent a lot of time in the desert, doing trail maintenance projects and fighting invasive species in the National Parks. "We did a lot of work with the Nature Conservancy, but we weren't politically engaged." But between his sophomore and junior year, he spent the summer in the Missouri Ozarks leading wilderness trips for teens, and realized this was what he wanted to do. So in the middle of his third year at ASU, Tim dropped out and moved to southern Missouri, where he worked with the YMCA doing outdoor education and leading trips into the mountains.

Tim moved to Utah in 2005, in large part because he wanted to work in a wilderness therapy program for teens that would be "more intense." The longest program in Missouri was only two weeks long. "I always felt by the end that we were just getting into the good stuff, they were just opening up and getting comfortable." So he went to work for a company called Outback, where so-called troubled kids were in the program for at least eight weeks.

Wilderness had long been an important part of Tim's life. Growing up, his family took vacations in West Virginia, New Hampshire, Montana, Wyoming. In his interview with Terry Tempest Williams, he recalled that when he was sixteen or seventeen, "struggling with all this teenage angst, and being overwhelmed with the world," his mother told him: "You need to go to the wilderness." She sent him down to the Otter Creek Wilderness in the Monongahela National Forest in West Virginia, where he'd been a number of times before. "I spent eight days there alone," he told Tempest Williams, "and it was a really powerful experience that led to my formation as an individual. I mean, it was the first time that I ever experienced myself without any other influences. . . . And it was terrifying to experience that." But it allowed him, he says, "to develop that individual identity of who I was without anyone else around." Working in Utah, he says in that interview, "I was fully into the wilderness . . . I was out there all the time. That's where I lived."

But Tim grew dissatisfied with the wilderness program, he told me, because he came to realize that these "troubled" kids were perfectly reasonable, and the program was only helping them adapt to "an unreasonable world." So in 2007, deciding that he needed to understand the structures underlying that unreasonable system, Tim went back to school, studying economics at the University of Utah in Salt Lake City—and that's when everything started to change.

That fall, just as the IPCC's *Fourth Assessment Report* was being released—a report that raised the alarm as never before, showing that climate change was arriving much sooner than predicted—he started paying close attention to what was happening to the planet. He joined a listserv on climate and environmental issues, run by one of his economics professors, where he connected with much of the Utah climate community. He got involved with the university's Office of Sustainability and with the Post Carbon Institute in Salt Lake.

Religion was also a big part of Tim's life, but in a new way. In Missouri, Tim told me, the community in which he had lived and worked was largely Christian, but he was no longer involved in a church. "I was thinking much more independently," he said, "studying independently." He read the Bible all the way through, but he also read Kierkegaard and Martin Buber and Tolstoy's religious writings, including *The Kingdom of God Is Within You*, a major influence on Gandhi. "I was exposed to these other religious ideas, and I realized that I didn't have nearly the degree of certainty that that evangelical Christian identity demanded. I felt like I was very much at the beginning of this religious exploration." By the time he left Missouri and moved to Utah, Tim said, "I no longer identified as a Christian, because my idea of a Christian was still very narrow." Now in Salt Lake City, the young woman he was dating brought him to the Unitarian Universalist church she attended. He felt an immediate connection, unlike anything he'd experienced before. "I remember the first time I went," he told me. "The sermon was quoting Emerson, maybe Thoreau as well, and I was like, 'Wow. That's amazing.'"

The congregation was also engaged on climate and social-justice issues. He got involved, and the church became the place where he first found his voice on climate, speaking especially from a generational perspective—something many of the Boomers in the congregation, who had children his age or younger, found compelling. "All the dire predictions," Tim recalled, "what's going to happen by 2030, 2040, 2050—I said, those aren't dates for me, they're ages. I knew how old I was going to be. And they hadn't thought of it that way before. They started saying, I know how old my kids are going to be on those dates."

The turning point—the moment, it's safe to say, that ultimately led Tim to prison and to the path he was now following—came in March 2008, when the University of Utah hosted a symposium on climate change. It was there that Tim met Terry Root, the IPCC climate scientist, whose interaction with him he has described as "shattering."

"She said it was too late," he recalled, to avoid devastating consequences of climate change.

"That same day," Tim told me, "after I had that experience with her, I met Bill McKibben. He was speaking the next day at the conference—he was going to announce 350.org—and I ran into him in the lobby of the hotel as I was leaving that afternoon. I was just some random kid. But I'd been blown away by Terry Root, and I was totally reeling from that, and then I see Bill McKibben in the hallway, and I just unloaded with him—all this emotion. And Bill was like, 'Yeah, I know. I got hit with that myself. But you've gotta get through it. I went through a really dark period, but I came through it afterward, and continued to keep pushing.' So he talked to me for a little bit."

"I stayed awake all night that night," Tim said. "There's only been two or three times in my life that I've stayed awake all night—I mean, I sleep like a rock, I slept soundly every night in prison. And as I was lying there awake that night, I thought, 'Well, all of my thoughts for my future are naïve and silly. They don't make any sense anymore. I've got nothing to lose by fighting back.' And I made a commitment to myself that I was going to say something the next day."

One of the speakers the following day was Dianne Nielson, energy advisor to Utah Governor Jon Huntsman. Tim had heard her speak previously to a group of students, and her message on climate change had been, as he recalled, "Don't worry, people in power have this covered." When a student had asked what young people could be doing, Tim remembered, "She told us to think about the little ways we can cut our emissions. She literally said, 'Turn off your lights and open the blinds.'"

When she got up to speak at the symposium the next day, Tim was ready. "She was talking about how the legislature and other decision makers really wanted to hear what we had to say, that it makes a big difference when you work through the system and file public comments, and all that bullshit. And during the Q&A I stood up and said, 'What is it going to take for them to actually hear what we have to say?' And I said, 'Just to clarify, I'm not asking for some advice about turning off the lights and opening the blinds. I want to know where and when do we need to take to the streets to finally get your attention?'"

"And the whole room just erupted," Tim said. "All these people were coming up to me afterwards, giving me their cards."

"And then Bill McKibben talked," Tim recalled. McKibben was introducing the concept of 350 parts per million, and the work of James Hansen arguing that 350ppm should be the upper limit of carbon dioxide in the atmosphere if humanity wanted to preserve a stable climate—and that we were already pushing 400ppm, with no end in sight. "And Bill said that he was starting this group called 350.org," Tim told me, "and they'd be coming out later with some clear requests. And then he goes, 'Like that kid just said, it's about when and where we need to take to the streets.'"

Tim began to study social-movement history and to engage all the more deeply with his church. He started thinking seriously about civil disobedience, "pushing the next boundary"—the need, as he told me, for a "more confrontational movement." And he began to realize, he said, that he couldn't put all his hopes on someone like Bill McKibben. "He's just a normal guy, and he's got no more power to make this happen than anybody else," Tim remembers thinking. "It was this sort of terrifying realization that I was my own best hope for my future. That nobody else was going to get this done for us."

During this time he also thought a lot about the seventh Unitarian Universalist principle, Tim told me. "Respecting and promoting the interdependent web of all existence, of which we are all a part—that we are not isolated individuals, but deeply interconnected. That resonated with me and empowered me."

That fall, as he was preparing himself to take some sort of bolder action—though precisely what sort, he didn't know—Tim learned that the Bureau of Land Management would be holding an auction in Salt Lake City in December, selling off the rights to drill for oil and gas on pristine public lands in southern Utah.

"The morning of the auction," Tim told me, "I had a final exam in a class called Current Economic Issues, or something like that, and one of the questions was: In this auction that's happening later today, if it's only oil and gas men bidding, will the final price reflect the true cost of developing oil?"

"And so I answered that question, which was all about the externalities," Tim said, "knowing that I was going to protest there."

When Tim arrived at the auction, there was a tame protest going on outside, people holding signs. He walked up to the entrance, but he had no plan for how to get inside, or what exactly he would do if he got in—make a speech, perhaps, force the police to drag him out. He fully expected to be turned away at the door.

"The security guard said, 'Are you here for the auction?' And I said, 'Yes, I am.' And he pointed me over to the table, and the official said, 'Are you here to be a bidder?' And I said, 'Well, yes. Yes, I am.'"

———

When Tim finally went to trial, after two years of delays, in the early spring of 2011, he and his lawyer had hoped to mount what's known as a "necessity defense." That is, Tim wanted to argue that his actions—bidding on and winning parcels for nearly $2 million he couldn't pay—were justified, and indeed necessary, given the imminent threat of catastrophic climate change. But the necessity defense is rarely allowed, and even more rarely successful. It was hardly surprising that the federal judge in Tim's case barred him from using it.

How fitting and strange, then, that on a Saturday afternoon in February 2014, two months after our conversation at Shay's, Tim and I found ourselves in a room together at the Quaker Meeting House on Beacon

Hill with Jay O'Hara and, via Internet connection, Ken Ward (who was living with his son in Corbett, Oregon), as well as others from the Boston-area climate movement. The topic was the necessity defense—Ken's and Jay's. The judge in the case of the Lobster Boat Blockade, to be tried in Bristol County court in Fall River, Massachusetts, had indicated that their defense could move forward, and the prosecution had presented no obstacles. It appeared that Ken and Jay would be the first climate activists ever to use the necessity defense in a US court. James Hansen and Bill McKibben had agreed to appear as expert witnesses. Matt Pawa, a leading environmental law expert and legal director of the Global Warming Legal Action Project, and Joan M. Fund, a top criminal defense attorney in Bristol County, would represent Ken and Jay. Those of us at the meeting were there to discuss how to organize and message around the trial, which had the potential to be a major event.

In order to mount a necessity defense in Massachusetts court, the defendant has to meet three requirements: first, that there was a "clear and imminent danger"; second, that there was a reasonable expectation that the action taken would be "effective in directly reducing or eliminating the danger"; and third, that there was "no legal alternative which would have been effective." It was the last one, in particular, that most appealed to Tim. It raised the very question, he pointed out, facing the climate movement as a whole: whether it's still possible to work effectively within the system, or whether our democracy is so corrupted, our government so complicit in its own takeover by corporate interests, that it's now necessary, given the scale and urgency of the threat, to work outside the rules as well—even if that means the risk of going to prison. If played right, it would put the system itself on trial.

I thought I understood where Tim was coming from. Just two nights earlier, he and I had sat down for another of our long conversations, at a packed and noisy restaurant in Harvard Square. He started out by telling me that he'd just given a speech that past weekend at Oregon State University, at a symposium called "Transformation without Apocalypse," in which he argued that the mainstream, NGO-led climate movement has failed—and, in fact, can only fail—and should therefore shut itself down and make way for a new and more honest kind of movement centered on climate justice and committed to challenging those in power.

"The mainstream climate movement has absolutely nothing to stand on when it comes to efficacy," Tim told me. "Nothing. They've had billions of dollars poured into these NGOs for decades, and they haven't done shit. I think being in the environmental movement for a long time should be considered a liability. It should be like someone who stands up and says, 'I've been in Congress for thirty years.' You know, you better have a good excuse."

I knew he was trying to be provocative—he admitted as much. In fact, he told me later that he came to regret aspects of his Oregon speech, for which he took a fair amount of heat. "I meant what I was saying to be taken *seriously*, but not *literally*," Tim told me. "Expanding what's possible to be discussed and considered, creating that space, can be healthy for the movement." In other words, sometimes you have to go to the edge, as he once said, and push.

Nevertheless, I pushed back. I understood the frustration and the anger toward the mainstream environmental movement, especially the Big Green NGOs and their failure to confront the true dimensions of the catastrophe. I felt it, and I'd talked with others, like Ken Ward, who felt it even more—who'd lived it. But I was skeptical of such a wholesale indictment—much less the notion that we should, or could, somehow replace the organizations, large or small, that formed the movement's infrastructure and that had the resources and the scale to meet the magnitude and urgency of the political challenge we faced. After all, those people are working hard, devoting their lives, to keep fossil fuels in the ground, still our overriding moral imperative if we're going to salvage any hope of climate justice, social justice, in the future. And in the near term, that means not only pushing from the outside. It often means working within the current political system, the only system we've got.

"But the kind of change you're talking about—anything feasible within the current political system—really won't do us any good," Tim shot back. "You're talking about going off the cliff at forty miles per hour instead of sixty."

"So, yes," he said, "the most urgent thing is keeping fossil fuels in the ground. The question is how to do that. We need a different kind of movement, a movement that's about taking power and changing power structures on a fundamental level. And I'm saying the climate movement is not equipped for that kind of struggle. The climate movement that has grown out of the environmental movement—primarily driven by comfortable people, rich people, white people—is about keeping things more or less the same. That's no longer the challenge that we have. On a really fundamental level it's about retaking power, and challenging those at the top—and the movement that we have has proven to be completely ineffective at that, and unwilling to take that challenge on."

The only places Tim saw that kind of movement being built, he said, were in the kinds of groups making up the Climate Justice Alliance, or Indigenous movements like Idle No More and the Moccasins on the Ground gatherings, the people on reservations in South Dakota fighting the "black snake," Keystone XL; or those groups in Appalachia fighting mountaintop removal.

"I don't think it's a coincidence," he said, "that it's the groups from impoverished and oppressed areas or oppressed constituencies that are building the kind of movement we need. I think it's because they've experienced part of the challenge that lies ahead for all of us—when there's plenty of reasons for hopelessness, they've chosen to fight back."

If our primary challenge now, in addition to the urgency of keeping carbon in the ground, is "how to maintain our humanity" in the face of inevitable hardship and scarcity, then it becomes all the more important, Tim is saying, that our movement have social justice at its core—and the willingness to confront power. "Looking only at the physical impacts, the food shortages, droughts, floods, mass migrations—that's really bad," Tim said. "But if you look at how our current power structure would deal with that—we have a power structure in our society, without a doubt, that is willing to scapegoat classes of people, pitting people against one another. And that's where things get really, really ugly."

"The determining factor in whether this crisis turns us toward one another or pits us against one another—whether it uplifts our humanity or kills our humanity—is how we go into it, and what kind of power is rising as we go into it. Whether it's corporate power or a people power, a community power."

Holding onto our humanity in the face of what's coming, Tim said, will be "a never-ending challenge." He quoted Alice Paul, one of the great leaders in the movement for women's suffrage, who famously said, "When you put your hand to the plow, you can't put it down until you get to the end of the row." Now, Tim said, "we're in a position where there's no end to the row. We need an endless movement and a constant revolution."

It would be several months before I sat down with Tim again. In early July, when classes were out, he invited me to the Divinity School. He wanted to know if I'd ever been inside Emerson Chapel, in Divinity Hall, where Ralph Waldo Emerson delivered his scandalous "Divinity School Address" (after which he was no longer welcome at Harvard). I had not. "Let's meet there," Tim suggested.

When I arrived, midmorning, we had the august room to ourselves. Less chapel than Puritan lecture hall, it's nevertheless a handsome, wood-paneled space, with large windows on one side looking out onto quiet Divinity Avenue and on the opposite wall portraits of ancient grandees, including the radical abolitionist Theodore Parker. A large plaque near the door informs visitors of the room's special place in the history of American religion, scene of Emerson's heretical speech. We could use a few more of those, I thought.

We sat on wooden chairs in the middle of the room.

So, do you believe in God? I asked.

"Yes," Tim said. "I don't claim to know the nature of that God, but yeah, I definitely believe that there is that force much greater than ourselves, much greater than we can understand."

Is it possible to describe it beyond that?

"I think there are ways to wax poetic beyond that."

Many people have tried, I said.

"I think all of those are explorations, trying to wrap ourselves around something that we fundamentally cannot understand. To me that's part of the nature of God."

"I believe that we have that divine presence within ourselves," he said. "I think it's that part of ourselves that is unlimited and that cannot be contained, that cannot be owned and controlled."

I remembered that Tim has called Jesus a "revolutionary figure," and I wondered if he now thought of himself as a Christian. Tim told me that while he was in prison, his minister in Utah had sent him the three-volume history of liberal religion by Gary Dorrien, now one of his professors at the Divinity School, which starts around 1800 with the birth of Unitarianism. "And looking at this history of liberal Christianity," Tim said, "I realized that the evangelical brand of Christianity that I'd been exposed to was a very narrow range of what is considered Christian, and a fairly modern interpretation of it, actually."

He'd been involved in the Unitarian Universalist church for several years, and at the church in Salt Lake City, he said, people were willing to abandon the Christian label. "But after studying religious history, I realized that people like me, who try to follow the example of Jesus—the message and example of Jesus is one of the really guiding things for me—I have at least as much right to call myself a Christian as the people who preach greed and violence and intolerance in the name of Jesus. So I tentatively started calling myself a Christian again, without being too attached to it as a label. If somebody asked, I'd say, yes, I'm a Christian."

It occurred to me that I hadn't asked Tim much about his experience in prison. It was easy to forget that the guy sitting across from me spent almost two years in federal custody. I asked how prison had affected him—if it had affected him spiritually. What he said surprised me.

"I think it was the first time I really believed in evil," Tim said. "I learned that there are institutions that are inherently, fundamentally evil—because they suppress the goodness in people, and they dehumanize them."

And then Tim told me a story. There was a guy who came in when he'd been in prison a few months. He was Latino, Tim said, and the guy had described a fairly common experience. He had a kid at a young age, got married. "And he can't find any real opportunities to provide for the kid and the family," Tim said, "and he ends up distributing, shuttling drugs for somebody. And so he gets caught, gets sentenced, and he gets ten years or whatever."

After he'd been in for a month or two, Tim said, the guy's wife and children came to visit him. There were two children, one who was about four and one who was maybe two, and when the visit was over and it came time to leave, and for the father to line up with the other inmates to be patted down, the younger child couldn't understand why he wasn't allowed to stay.

"He hadn't seen his dad in months," Tim said, "and he just couldn't understand why his dad couldn't come with them. They had to drag the kid away. He was screaming and crying—and the guard was like, *'Well, he's gotta go. He's gotta go right now.'* And the kid's clinging to his dad's leg, and the mom is pulling him off, and she's crying, like, *'What should I do?'* And the father's like, *'Just get him out, get him out.'* And the father's crying, like, openly crying. And all the other inmates are crying. All of us have tears in our eyes. I mean, inmates don't go around crying in front of other inmates, and all of us were crying, choking back tears—I still can't even tell the story."

He stopped, and the chapel was suddenly very quiet. I looked up, and realized Tim's eyes were watery. He cleared his throat, resumed the story.

"The guard had no reaction. I mean, there are guards who are real sadistic, disturbed human beings. This one wasn't. But he had no empathy. It didn't faze him. And I thought about what it would take to do that job, on a daily basis, and how the only way you could continue doing that job is to dehumanize these people—and to shut down that empathy within yourself, and to be dehumanized, in order to continue that cycle."

"It's a fundamentally evil system," he said.

I asked Tim if he had suffered, personally, while he was in prison.

He looked at me.

"Only emotionally."

The next time I saw Tim at Harvard Divinity School, months later, he was speaking at an interfaith conference on climate change and religion, not in Emerson Chapel but a modern lecture room in Andover Hall, with a PA system and stadium-style seats. He was the final speaker, and the only student, on a panel that included Buddhist, Jewish, Muslim, and Christian faith leaders, and he addressed an audience that included faculty and alumni and interested community members like me. He was in fighting form, and standing behind the lectern he spoke passionately, warmly, with humanity.

He reminded us of the magnitude of the challenge we face, and what's at stake in the fight for climate justice—that we intend to cost "the richest and most powerful and most ruthless industry in the world," he said, "trillions of dollars in lost future profits." What's more, he said, "this is an industry that has killed people for profit throughout its whole existence." In his home state of West Virginia, he said, people have been dying because of fossil-fuel extraction since the industry began. And with new, experimental, more destructive and toxic forms of extraction, such as fracking wells and tar-sands mines, even more people are dying all across the continent. "And since it's rather hard to convince people to poison their own water and blow up their own backyards," Tim said, "the fossil-fuel industry has increasingly moved towards killing democracy in order to be able to continue killing people for profit."

"This is our challenge," Tim said. "To overcome an industry that has made it clear that they are not going to willingly get out of the way and join us in a quest for a healthy and just future. So if we're going to stand in the way of that kind of systemic evil, we better be standing on pretty solid rock." He asked if faith communities, like those represented in that room, have what it takes to be that rock.

"Because it's not 1992 anymore," he reminded us. "It's 2014. And we haven't been doing what we needed to be doing for the past generation."

"Even if we're as successful as we can possibly be," Tim said, "there will be massive impacts and unprecedented hardship in our future, probably at catastrophic levels. And if history is any guide, as we go down that path of chaos and desperation, we can expect our leaders to do desperate things to try to cling to their power."

"And if our past social movements are any indication," Tim said, "we know that there will be points where there is no reasonable expectation that we will win this struggle. And in those dark moments, we will have to continue the struggle for justice, not because we expect that things will be OK—in those dark moments, we will continue the struggle for climate justice because that's what it means to be faithful to a God who loves this world. We will continue this struggle because that is what it means to be faithful to the people and to the world that we love."

———

What are we fighting for? What are any of us who care about climate justice really fighting for? What does "climate justice" mean in the face of the inhuman and dehumanizing maw of the world-devouring carbon-industrial machine—of which we ourselves are a part? What does it mean in the face of the science—which keeps telling us, in its bloodless language, just how late the hour really is?

It seems that movements often reach a critical juncture at which unity—the need to come together around common principles and a common struggle, and a common understanding of what that struggle is about—becomes all-important. Or if not unity—which may in fact be impossible for any movement big enough and broad enough to be powerful—then at least something like solidarity. So I ask again: At this late hour, what are we fighting for?

Trust me, I know full well that any talk of a "transformative," "radical" movement for climate justice, or any kind of deep political transformation, sounds hopelessly naïve. I get it. I know. I know the country, and the political culture, in which I live and work.

And yet—here I am anyway. Because I also know that abolishing slavery sounded hopeless and naïve in 1857, when Frederick Douglass spoke of struggle. I know that throwing off the British Raj sounded hopeless and naïve in 1915, when Gandhi returned to India. I know that ending Jim Crow sounded hopeless and naïve in 1955, when Rosa Parks stayed in her seat on that bus in Montgomery. I know that ending apartheid sounded hopeless and naïve in 1962, when Nelson Mandela went to prison in South Africa.

For that matter, even stopping the Keystone XL pipeline sounded hopeless and naïve in 2011—before thousands of people started getting arrested and literally laying their bodies on the line, with tens of thousands more pledging to do so, in order to stop it. And before a president of the United States started listening. Yes, the southern leg got built. And yes, the whole thing is just one pipeline—one very big, very dangerous, very symbolic and political pipeline. And yes, Montgomery, Alabama, was just one Southern city. And that bus was just one city bus.

And so history says never quit, never give up. But science—we keep coming back to the science. And the science keeps telling us just how late it is. And it's true—we have to be honest, with ourselves and with others. After all, what good is a movement if it can't even be honest about the situation it faces? We have to ask ourselves, in all honesty, given the facts we now face, what it is that we're really fighting for.

On April 4, 1967, the Reverend Dr. Martin Luther King Jr. rose to speak from the pulpit of the Riverside Church in Manhattan's Morningside Heights—a church built by John D. Rockefeller, the founder of Standard Oil—and delivered what was perhaps the boldest, most radical speech of his too-short life. Daring to denounce the war in Vietnam as a national sin eating away at the American soul, and prophetically proclaim the war's inseparability from the struggles for racial and economic justice, King knew that he would alienate, maybe even lose, some of his strong allies—many of whom were not yet willing to break with Lyndon Johnson and the still prowar liberal establishment. People would ask him, he noted, "Why are you speaking about the war, Dr. King? . . . Peace and civil rights don't mix."

When he heard this, King said that day, he was "greatly saddened, for such questions mean that the inquirers have not really known me, my commitment, or my calling." That commitment and calling, he reminded them, was that of a Christian preacher. "Have they forgotten that my ministry is in obedience to the one who loved his enemies so fully that he died for them?"

King had just written his final book, *Where Do We Go from Here: Chaos or Community?*, to be published that June. And in those pages, as in his speeches during those last two years, King struggled to reinvigorate and reunite the civil rights movement, which was coming apart at the seams over Black Power and nonviolence, separatism and integration, and over how fast and how hard to push for economic justice and against the war in Vietnam. And while he's often cast these days as a soothing moderate, it's important to remember just how radical King was, especially at the end of his life. Establishment critics, including the NAACP, thought that he should keep his focus on race and civil rights and not stick his nose into what they considered the "separate issues" of labor, poverty, and, most of all, the war. But King understood that all of these issues were, at a profound level, interconnected—he saw their *intersectionality*. He knew, as he wrote in a Birmingham jail cell in 1963, that "injustice anywhere is a threat to justice everywhere." He saw systemic evil, and knew it required a systemic solution. He saw, and argued forcefully, that the "unholy trinity" of racism, poverty, and war were, at root, one and the same—that they are all forms of violence, that they all grow from "man's inhumanity to man," and can be defeated only by universal love.

"A genuine revolution of values," King declared at Riverside Church, in a passage drawn from the final pages of his new book, "means in the final analysis that our loyalties must become ecumenical rather than sectional. Every nation must now develop an overriding loyalty to mankind as a whole in order to preserve the best in their individual societies."

"This call for a worldwide fellowship that lifts neighborly concern beyond one's tribe, race, class, and nation," he continued, "is in reality a call for an all-embracing and unconditional love for all mankind. This oft misunderstood, this oft misinterpreted concept, so readily dismissed by the Nietzsches of the world as a weak and cowardly force, has now become an absolute necessity for the survival of man. . . . I am speaking of that force which all of the great religions have seen as the supreme unifying principle of life. Love is somehow the key that unlocks the door which leads to ultimate reality."

At which point, the Baptist preacher quoted from the first epistle of Saint John, chapter four: "Let us love one another: for love is of God."

Of course it's easy for me to quote Martin Luther King, and feel good about myself for doing so. But there's nothing easy about the path he showed us and the gospel he preached. As with Thoreau, we who invoke

King need to ask ourselves if we're ready to walk in his footsteps. King was a radical and a revolutionary—ready to *give his life* in the cause of justice. When those of us who appropriate his words can say the same, then maybe we can claim some tiny portion of his legacy.

The title of King's last book poses the question facing us now: Where do we go from here—chaos or community? We know that the crimes perpetrated against the planet today are a form of violence against our fellow human beings, a profound racial and economic and generational injustice, on a global scale. And we know that if we're going to have a movement big enough and powerful enough to confront this, we have to come together across our cultural, racial, economic, and generational divides, even our ideological divides.

Nothing about that task is easy. Indeed, if there's to be any hope of such solidarity, then "climate justice" will need to be defined broadly enough, inclusively enough, to encompass everyone—not only certain communities, not only *our own* communities, and our own children, but everyone, everywhere, including generations not yet born—in order to keep even the possibility of justice alive on Earth.

Because what we're fighting for now is each other. We have to fight for the person sitting next to us and the person living next door to us, for the person across town and across the tracks from us, and for the person across the continent and across the ocean from us. Because we're fighting for our humanity. That's what solidarity is. That's what love looks like. Not simply a fight for our own survival—there are oppressive and dystopian forms of survival that aren't worth fighting for. Indeed, that are worth fighting *against*. Rather, ours is now a fight for survival and a fight for justice—no, for the survival of the *possibility* of justice and some *legitimate hope* for what King called the "beloved community." Even as we struggle just to survive.

Our fight is against chaos—and for community. And it cannot wait.

There at Riverside Church, King spoke words that would appear in the final paragraph of that final book: "We are now faced with the fact that tomorrow is today. We are confronted with the fierce urgency of *now*. In this unfolding conundrum of life and history there is such a thing as being too late. . . . Over the bleached bones and jumbled residues of numerous civilizations are written the pathetic words: 'Too late.'"

Is it too late? We know what the science says. What does your conscience say? What does "too late" even mean? Too late *for what*? Even in the face of all we now know, will it ever be too late for some kind of *faith* in human decency; or to hold onto some kind of *hope*, however irrational it may seem, in our fellow human beings; or to *love* our brothers and sisters on this earth?

Because these things—faith and hope and love—are every bit as real as the science, every bit as real as the CO_2 in the atmosphere and the carbon in the ground. As real as the melting Arctic and the acidifying oceans. And as Dr. King knew, these things—faith, hope, love—are the very stuff that movements are made of: real movements, the kind of radical, transformative movements that have changed the course of history in the past, and maybe, just maybe, might change it again. If enough of us are willing to fight, and to fight hard enough, and to fight lovingly enough, and never give up. If we're willing to engage in this struggle—this radical and loving struggle—for each other.

———

In one of our conversations that winter, I confessed to Tim that I'd been seriously tempted to join Ken and Jay on the lobster boat at Brayton Point—but that I'd struggled, hard, with what it could mean for my family, for my wife, my children. I feared losing them. I wasn't ready to risk being taken away from my kids, I told him. Having a family, having children, makes the moral calculus more difficult. My children were the first reason I got into this fight, I said, and I couldn't bring myself to make them suffer.

"There's a big difference between a moral calculus and morality," Tim said.

If I had a community, I said, that I knew would hold them, take care of them, it would make it easier.

"That's why we need the kind of movement that really values relationships," Tim said. "A community that can make sense of this for them—that can say, this is why your dad went to jail."

And Yet

A thousand grasses bend with dew,
A hill of pines hums in the wind.
And now I've lost the shortcut home . . .

    —GARY SNYDER, "COLD MOUNTAIN POEMS," 1958

Almost swallowed by the vastness of the mountains,
but not yet.

As the barely audible
geese are not yet swallowed;
as even we, my love, will not entirely be lost.

    —JANE HIRSHFIELD, "THE HEART'S COUNTING KNOWS ONLY ONE," 1997

Henry's jaw would have dropped. On the morning of September 8, 2014, in Fall River, Massachusetts, something truly remarkable, a kind of blessed unrest, took place: for a moment, a higher law, the law of conscience, held sway in the commonwealth.

It happened at the Bristol County courthouse, where Ken Ward and Jay O'Hara were going to trial for their blockade of the coal freighter at Brayton Point. They would be the first in the United States to use a "necessity defense" in a civil-disobedience trial centered on climate change. James Hansen and Bill McKibben, among others, were lined up as expert witnesses. Hundreds of supporters had gathered at the courthouse, where an hour before the doors opened dozens of Quakers, friends of Jay's and Ken's, joined in silent witness on the public plaza.

Inside, the courtroom quickly filled beyond capacity. In the front row, just behind Ken and Jay and their lawyers, Tim DeChristopher—one of a team of organizers, myself included, who had prepared to publicize the trial—took notes and handed them to a runner to be sent as updates via social media. Before too long, the court recessed, and Ken and Jay huddled with Tim over the railing. Tim pulled me aside, asked if I knew the reporters who were in the room, because we were about to make some news.

And then what happened, the truly remarkable thing, was this: the prosecutor, Bristol County District Attorney Sam Sutter, dropped the charges—which could have resulted in months, or even years, of jail time. But that's not all. He then proceeded out to the courthouse plaza, where he made a statement to the media and the hundreds of people gathered. Ken, Jay, and Tim stood nearby, along with the defense attorneys and an interfaith contingent of clergy.

"The decision that Assistant District Attorney Robert Kidd and I reached today was a decision that certainly took into consideration the cost to the taxpayers in Somerset," Sutter said to the cameras, "but was also made with our concerns for their children, and the children of Bristol County and beyond in mind."

There was something in Sutter's tone, a kind of earnestness one doesn't often hear, and a rush of emotion, like an electric current, passed through the crowd pressing in around the microphones. I stood just a few feet in front of the DA, and realized a significant lump had formed in my throat.

"Climate change is one of the gravest crises our planet has ever faced," Sutter said, his voice clear and resolute. This was a speech, not a statement. "In my humble opinion, the political leadership on this issue has been gravely lacking. I am heartened that we were able to forge an agreement that both parties were pleased with and that appeared to satisfy the police and those here in sympathy with the individuals who were charged."

The crowd went wild. A good many teared up. When the cheering settled down, a voice called out: "Will you be a model for across the country?" It was a friend of mine, as it happens, from Wayland: Rabbi

Katy Allen, a cofounder of Transition Wayland and the Boston-area Jewish Climate Action Network.

"Well, I certainly will be in New York in two weeks," Sutter replied, referring to the much-anticipated People's Climate March on September 21. "How's that?"

The crowd thought that was pretty swell, too.

He added: "I've been carrying around Bill McKibben's article in *Rolling Stone*"—and he brandished the magazine with Bill's latest piece, a manifesto aimed at mobilizing the march.

OK, this guy's running for office, I thought. Fine with me. (Soon after, in fact, Sutter announced his candidacy for mayor of Fall River, a proud working-class immigrant city, and won the election that November.)

A reporter then asked if Sutter was sending a message condoning actions that violate the law. No, Sutter said, that's not the message. "I'm sending a message that this was an act of civil disobedience, that we had to reach an agreement. I'm not at all disputing that the individuals were charged, but this was the right disposition."

"Just to be clear," the reporter asked, "what would you say if people say in fact you're encouraging others to blockade tankers?"

"This is one case, one incident, at a time." Sutter responded. "I think I've made my position very clear. This is one of the gravest crises the planet has ever faced. The evidence is overwhelming, and it keeps getting worse. So we took a stand here today."

And so—sometimes we win. A small victory, perhaps, in the scale of the climate. But real. Something was happening here.

Meanwhile, as Ken and Jay were quick to point out, the Brayton Point plant burned on. The supposed 2017 closure, announced a few months after their action and the protests it inspired, wasn't nearly soon enough—for the climate, or for the plant's neighbors suffering its pollution.

What's more, as Ken noted to me, according to data from the US Energy Information Administration, the Brayton Point plant had essentially *doubled* its coal consumption in the previous year, and reduced its use of natural gas, making it still the first or second largest source of carbon emissions in New England, New York, and New Jersey. Observed shipments of coal to the plant had increased over the previous year.

"We have to shut it down," Ken said in May 2013—words that remained as true and as urgent as they'd ever been.

Later that fall, Ken, Jay, Tim, and their friend Marla Marcum, a core member of Ken's and Jay's support team, decided to create the Climate Disobedience Center, an initiative to encourage and support other activists who find themselves ready and willing to take nonviolent action commensurate with the climate catastrophe. They asked me if I'd join them—and I said yes. Yes, I would.

That bright May morning on the dock in Newport, Ken asked Marla, who had been at his and Jay's side from the start, if she would say a prayer. Marla, you may recall, was one of the cofounders of Better Future Project and 350 Massachusetts, and she's a good friend of mine. Marla reminds me of my own sisters. She grew up on a small farm in the Ozarks of southern Missouri, where her family struggled, and she was raised in a strong Methodist congregation that she credits with teaching her the essentials of community organizing. She went on to train for the United Methodist clergy, but before she could be ordained she took the leap into climate activism.

"The boat was ready to go, and we were all standing there looking at each other, wondering how to launch the next phase of this journey," Marla later wrote in an e-mail. "My two dear friends were putting themselves on the line that day, and they asked me to lead them in prayer." Marla said she wished she had a Bible, and Ken produced one from inside the boat. "To me," Marla wrote, "this was a very good sign."

Standing there on the dock with just Ken, Jay, and one other member of their support team, Ben Thompson of the Westborough Eight, Marla offered a prayer of dedication. "We give you thanks this day," she prayed, "for your call to love the world and to change it." She prayed "that the gifts we offer this day transform the hearts of many just as our own hearts are being transformed."

Marla opened the Bible to the fourteenth chapter of the Gospel of John, where Jesus speaks to his Apostles at the end of the Last Supper. "Peace I leave with you; my peace I give to you," she read. "I do not give to you as the world gives. Do not let your hearts be troubled, and do not let them be afraid."

She then turned to the first epistle of Saint John, the fourth chapter. And there, following the same passage recited by Dr. King at Riverside Church, Marla read: "Perfect love casts out fear."

———

One day in November 2014, craving a clear head, I went for a walk in the Great Meadows National Wildlife Refuge, at the Weir Hill headquarters on the Sudbury River, the snaking ribbon of water that flows north into the Concord River not far from Walden Pond. Weir Hill, south of Concord in Sudbury, is about three miles as the crow flies from my house on the Wayland side of the river. It's a place that became important to me soon after I started walking in 2006.

Henry Thoreau, on the opening page of his first book, *A Week on the Concord and Merrimack Rivers*, explains that before the Concord River was named by English settlers in 1635, it was known to Native Americans as the Musketaquid, or Grass-ground. "To an extinct race it was grass-ground," he writes, "where they hunted and fished, and it still is perennial grass-ground to Concord farmers, who own the Great Meadows, and get the hay from year to year."

"Between Sudbury and Wayland," Thoreau writes, "the meadows acquire their greatest breadth, and when covered with water, they form a handsome chain of shallow vernal lakes, resorted to by numerous gulls and ducks. Just above Sherman's Bridge, between these towns, is the largest expanse . . ."

Just downstream from Sherman's Bridge Road, and its present bridge, Weir Hill rises perhaps a hundred feet above the river. The hill, if you can call it that, gets its name from the fishing weirs used by the original human inhabitants along these banks. Thousands of relics of the early Indigenous people, dating as far back as 5500 BCE, have been found in the area of the Great Meadows. In the early nineteenth century, dams were built in the mill towns downriver, to the north, and the water extended into the grasses, creating a wetland habitat ideal for waterfowl. In 1944, a local landowner named Samuel Hoar donated 250 acres of the marsh-lands to the US Fish and Wildlife Service. The refuge now incorporates more than 3,800 acres along the Sudbury and Concord Rivers.

Henry, in those opening pages, describes "ducks by the hundred . . . gulls wheeling overhead, muskrats swimming for dear life . . . and count-less mice and moles and winged titmice along the sunny, windy shore; cranberries tossed on the waves and heaving up on the beach, their little red skiffs beating about among the alders—such healthy natural tumult as proves the last day is not yet at hand."

It was late morning, a brisk, clear day, the mercury stuck in the thir-ties—New England felt a lot like New England should. Frost came late again, and I was greedy for the biting air on my face and ears, the cold wind tearing my eyes, spotting my glasses. I stopped at the water's edge, at the foot of the hill, where the river bows around a small wooded pen-insula. The yellows and browns of the tall marsh grasses stretched for miles beneath a cloudless sky. Above my head was a blue so deep it hurt to stare into it for very long. The sun on the surface of the slow-moving water blinded me. Nothing stirred. It was quiet except for the breeze in the half-bare branches.

I walked the short, steep trail up Weir Hill, topped with oaks and white pines, and saw the river below through the trees, conscious that human beings, whose hearts pulsed like mine, lived and died where I stepped—that they walked and hunted and fished along that river—for seven thousand years. They knew that their lives were inseparable from the land, the water, the air—their lives, and their children's lives, and their children's children's lives. Forever. They knew. They knew.

In the spring of 2012, Gary Snyder—the great American poet and essay-ist whose "Cold Mountain" translations I quoted near the outset of this book—received the Henry David Thoreau Prize for "nature writing" from the organization PEN New England (as if Snyder, or Thoreau, can be bound by any genre), and I went to see him receive the award in a lecture hall at MIT, of all places—and, especially, to hear him read his poems. Spry, wry, and jovial at eighty-one, his white-bearded face was still elfin, if craggy and a bit gnarled, like a mountain or an ancient tree.

Snyder, old traveling buddy of Allen Ginsberg and Jack Kerouac, has long been celebrated as a poet and essayist of place—of Cascade peaks, Kyoto temples, Beat San Francisco, the South Yuba watershed in the Sierra Nevada foothills where he's made a life with his family since 1970. The idea of truly inhabiting one's surrounding landscape is vital to his ecological ethic, a kind of bioregionalism much influenced by Native American culture and spirituality.

But these days, just as much, I think of Snyder as our preeminent poet of time, impermanence, transience: of cosmic kalpas, geologic eons, and the evanescent ripple of the present moment.

Snyder, who was born and raised in Washington state and has lived most of his life on the West Coast, was attracted early to East Asian poetry and to Buddhism, with years of formal Zen training in Japan, so maybe his interest in transience is de rigueur. And sure enough, among the poems Snyder read at MIT that night were those highly regarded translations from the Chinese, made as a grad student at Berkeley in the mid-fifties, of the quasi-mythical Tang dynasty hermit Han-Shan, Cold Mountain. (Snyder mentioned, bemused, that in China he's been called "the American Han-Shan.") As though by request, Snyder's lilting baritone gave me back the lines that had run through my mind on those early walks to Walden and Great Meadows:

I settled at Cold Mountain long ago,
Already it seems like years and years. . . .
Happy with a stone underhead
Let heaven and earth go about their changes.

Hearing those lines, I thought of Thoreau on Katahdin—and how different this was from his visceral awe of "vast, Titanic, inhuman nature."

I was reminded, also, of the bristlecone pines—those trees as old as civilizations, some dating almost five thousand years, growing at high elevation on the rim of the Great Basin. And I remembered those reports, two years earlier, that the ancient trees are now threatened by climate change.

I'm sure those trees came to mind, too, because of a Snyder poem that has deeply affected me. In "The Mountain Spirit"—the climactic poem of his epic, decades-in-the-making volume, *Mountains and Rivers Without End* (1996)—Snyder tells of making a trip to "the angled granite face / of the east Sierra front" to sleep out among the bristlecones: "to the grove at timberline / where the oldest living beings / thrive on rock and air." There, as meteors streak the night sky, he meets the Mountain Spirit, at first an unseen voice and then an old woman with "white ragged hair." In his waking dream-encounter with the Mountain Spirit muse-mother (herself an amalgam, Snyder explains in a note, of traditional Japanese No play and Native American Ghost Dance), it's as though he hears the voice of geologic time, nonhuman time, and the inconceivable forces that formed the Great Basin:

> Ten million years ago an ocean floor
> glides like a snake beneath the continent crunching up
> old seabed till it's high as alps.

Deep in the night, the spirit asks him to read one of his old poems, which he calls "The Mountain Spirit" (a poem within the poem), its refrain echoing the final lines from yet another earlier poem, "Endless Streams and Mountains," the very poem with which the whole epic volume began:

> *Walking on walking,*
> *Under foot   earth turns*
> *Streams and mountains never stay the same.*

Geologic time, earth time. Snyder confronts it all with a kind of equanimity, a playfulness, even joy. "Ghosts of lost landscapes / . . . *Goose flocks / crane flocks / Lake Lahontan come again!*" he writes, invoking the name of that inland sea that once covered the Great Basin. He and the spirit mother dance among the pines: "The Mountain Spirit and me // like ripples of the Cambrian Sea // dance the pine tree." The earth will go about its changes:

> *Ceaseless wheel of lives*
> *red sandstone and white dolomite.*

Far from Thoreau's existential angst above the wind-battered timberline, there's a kind of transcendent calm at the center of Snyder's poetry. But what of the bristlecone? What of us? How to read Snyder in the face of the Anthropocene—in the face of the sixth mass extinction of living species since life on Earth began?

Head for the hills? Go climb a mountain?

Well, sure—why not? But unlike Han-Shan, and like Thoreau, Snyder is no hermit. He does not turn his back on the world. He never settles into passivity or fatalism. For decades, he has entered the fray. His Zen is nothing if not *engaged*—it leads him back down, off the mountain, among his fellow human beings. The path he would follow is the bodhisattva way.

I'm reminded of another Snyder poem. In "After Bamiyan," a prose-verse reminiscence from his 2004 collection, *Danger on Peaks*, Snyder recalls the destruction, in March 2001, of the ancient Buddhas of Bamiyan—carved into a mountain in modern-day Afghanistan—and records his exchange with a correspondent who noted smugly that Buddhism, after all, teaches the impermanence of all things.

"Ah yes . . . impermanence," Snyder replies. "But this is never a reason to let compassion and focus slide, or to pass off the sufferings of others because they are merely impermanent beings." He then quotes a famous haiku by the Japanese poet Issa:

Tsuyu no yo wa   tsuyu no yo nagara   sarinagara

Which he translates:

This dewdrop world
is but a dewdrop world
and yet—

Snyder adds:

That *'and yet'* is our perennial practice.

I keep those words close, as heaven and earth go about their changes—at a pace, and with a push from humans, never dreamed among those age-old pines.

———

At the edge of the woods along the Great Meadows: bright late-morning light, small birds on bare maples. A squirrel rustles, unseen, in dry leaves. A jet engine streaks overhead, and is gone. Calls of geese on the wind, somewhere over my shoulder, fade. A hawk circles high above the expanse of grasses, the snaking river, in the sun. White wisp of cloud, the only one, dissolves into deepest blue. The sun slants through pines.

No sanctuary. No refuge. It has only begun. We have only begun.

"Only that day dawns to which we are awake."

Wake up. Start again.

The sun climbs the sky.

ACKNOWLEDGMENTS

From early 2012 to mid-2014, I conducted more than a hundred interviews with more than seventy individuals involved in climate and environmental-justice activism. And although only a fraction of those interviews are quoted in these pages, each and every one of the people who spoke with me—and many others alongside of whom I've organized, marched, and engaged in nonviolent direct action—made an essential contribution to my understanding of the climate movement. Their stories and voices, whether they appear in these pages or not, are at the heart of this book—and are far more important than my own. I am profoundly indebted to all of them, and profoundly grateful.

The substance of this book grew out of essays, interviews, and features written from the fall of 2010 to the fall of 2014—all conceived as part of a single project. That previously published material has been significantly revised and updated, and in many cases greatly expanded. I am deeply grateful to the editors who encouraged, supported, and edited the articles from which this book grew: John Swansburg at *Slate*, Carly Carioli at the *Boston Phoenix*, Scott Rosenberg at *Grist*, Nicole Lamy and Paul Makashima at the *Boston Globe*, and Professor Susan Gallagher at UMass Lowell and the Thoreau Society. Most of all, I thank my editors at the *Nation*: Katrina vanden Heuvel, whose immediate grasp of my project, and unwavering commitment to it, came at a decisive moment; Roane Carey, whose patient guidance and keen editing shaped every piece for the magazine; and everyone whose hard work and dedication continue to make that magazine, founded by abolitionists 150 years ago, a pillar of independent journalism.

There are others whose insights and encouragement were invaluable. Bill McKibben, Alan Jacobs, Jenny Schuessler, Scott Stossel, and Tina Bennett generously read an early draft of "Walking Home from Walden," and gave me the confidence to keep writing. Mark Hertsgaard and Naomi Klein, before we became colleagues, saw something in my *Phoenix* essay "The New Abolitionists" in February 2013, and helped bring me into the *Nation*'s pages; I'm eternally grateful for their encouragement, friendship, and comradeship in this struggle.

Many, many people contributed to this book, but in the end there are two without whom it simply would not have happened.

My editor and friend, Alexis Rizzuto, believed in this project at a moment, in mid-2012, when I had all but given up on it. This really is Alexis's book as much as mine (though I alone bear the blame for its shortcomings). And of course Alexis is part of a remarkable and dedicated team at Beacon Press—including Beacon's visionary director, Helene Atwan—who made my journey into a book. Alexis, and all of her colleagues, have my deepest respect and gratitude.

But more than anyone, the person who made this work possible is my wife, Fiona—my only love and best friend since we were undergraduates on the Charles River, and the mother of our two beautiful children. Fiona supported me, in every sense, through four and a half years of what often seemed like wandering in the wilderness (or at least the trails on the way to Walden Pond). She joins me in giving our share of the proceeds from the sales of this book to grassroots climate and environmental-justice groups. Because she knows this book is nothing if not a labor of love.

My parents and sisters taught me what love is. And my children, Duncan and Grace, without ever realizing it, have taught me to love in ways I never knew were possible. My love is far from perfect. I'm only human. And so, for now, I give them this imperfect book. It's for them, and for all of their generation, that I wrote it—so that someday, when the going really gets rough, they will know that there were people in this country who were willing to fight for them, and their children after them, and all humanity.

ACTION RESOURCES

While this is far from a comprehensive list of movement organizations, please consider supporting and/or learning how to get involved with the groups and networks discussed in this book.

350.org / FossilFree
   gofossilfree.org

Alternatives for Community and Environment (ACE), Boston, MA
   www.ace-ej.org

Asian Pacific Environmental Network (APEN), Oakland, CA
   apen4ej.org

Better Future Project / 350 Massachusetts, Cambridge, MA
   www.betterfutureproject.org
   350MA.org

Black Mesa Water Coalition, Black Mesa, AZ
   blackmesawatercoalition.org

Climate Disobedience Center
   climatedisobedience.org

Climate Education Community University Partnership (CECUP), Barbara Jordan-Mickey Leland School of Public Affairs, Texas Southern University, Houston, TX

Climate Justice Alliance: Our Power Campaign
www.ourpowercampaign.org

Communities for a Better Environment, Oakland and Huntington Park, CA
www.cbecal.org

Community In-Power & Development Association (CIDA), Port Arthur, TX

Deep South Center for Environmental Justice, Dillard University, New Orleans
dscej.org

Dudley Street Neighborhood Initiative, Boston, MA
dsni.org

East Michigan Environmental Action Council, Detroit, MI
www.emeac.org

Idle No More, North America
www.idlenomore.ca

Indigenous Environmental Network, North America
www.ienearth.org

Kentuckians for the Commonwealth, London, KY
kftc.org

Michigan Coalition Against Tar Sands (MI-CATS)
www.michigancats.org

Mobile Environmental Justice Action Coalition, Mobile, AL
facebook.com/mejacoalition

Mothers Out Front: Mobilizing for a Livable Climate
mothersoutfront.org

NAACP Climate and Environmental Justice Initiative
www.naacp.org/programs/entry/climate-justice

New Economy Coalition
  neweconomy.net

Peaceful Uprising, Salt Lake City, UT
  www.peacefuluprising.org

Radical Action for Mountain People's Survival (RAMPS), West
Virginia
  rampscampaign.org

Resilient Nacogdoches (formerly NacSTOP), Nacogdoches, TX

Rising Tide North America
  risingtidenorthamerica.org

Students for a Just and Stable Future (SJSF), New England
  facebook.com/JustAndStable

Tar Sands Blockade
  www.tarsandsblockade.org

Texas Environmental Justice Advocacy Services (TEJAS), Houston, TX
  tejasbarrios.org

SELECTED BIBLIOGRAPHY

Including sources referenced in the text and relevant works consulted.

Alperovitz, Gar. *America Beyond Capitalism: Reclaiming Our Wealth, Our Liberty, and Our Democracy.* Hoboken, NJ: John Wiley & Sons, 2005.

———. *What Then Must We Do? Straight Talk About the Next American Revolution.* White River Junction, VT: Chelsea Green, 2013.

American Friends Service Committee. *Speak Truth to Power: A Quaker Search for an Alternative to Violence.* Philadelphia: American Friends Service Committee, 1955. http://www.afsc.org/sites/afsc.civicactions.net/files/documents/Speak_Truth_to_Power.pdf.

Anderson, Kevin. "Climate Change Going Beyond Dangerous—Brutal Numbers and Tenuous Hope." *Development Dialogue* 61 (September 2012): 16–40.

———. "Climate Change: Going Beyond Dangerous." Slide show presented at the Department for International Development, London, July 2011. http://www.slideshare.net/DFID/professor-kevin-anderson-climate-change-going-beyond-dangerous.

———. "The Emission Case for a Radical Plan." Slide show presented at the Radical Emissions Reduction Conference, Tyndall Centre for Climate Change Research, Norwich, UK, December 11, 2013. http://www.tyndall.ac.uk/sites/default/files/anderson_-_radical_plan_conf.pdf.

Athanasiou, Tom, and Paul Baer. *Dead Heat: Global Justice and Global Warming.* New York: Seven Stories Press, 2002.

Berry, Wendell. *The Art of the Commonplace: The Agrarian Essays of Wendell Berry*. Edited by Norman Wirzba. Berkeley, CA: Counterpoint, 2002.

———. "Christianity and the Survival of Creation." In Wirzba, *Art of the Commonplace*, 305–20.

———. "The Gift of Good Land." In Wirzba, *Art of the Commonplace*, 293–304.

———. *It All Turns on Affection: The Jefferson Lecture and Other Essays*. Berkeley, CA: Counterpoint, 2012.

———. *New Collected Poems*. Berkeley, CA: Counterpoint, 2012.

Bigelow, Albert. *The Voyage of the Golden Rule: An Experiment with Truth*. New York: Doubleday, 1959.

Black, Toban, Stephen Darcy, Tony Weis, and Joshua Kahn Russel, eds. *A Line in the Tar Sands: Struggles for Environmental Justice*. Oakland, CA: PM Press, 2014.

Branch, Taylor. *At Canaan's Edge: America in the King Years, 1965–68*. New York: Simon & Schuster, 2006.

———. *Parting the Waters: America in the King Years, 1954–63*. New York: Simon & Schuster, 1988.

———. *Pillar of Fire: America in the King Years, 1963–65*. New York: Simon & Schuster, 1998.

Brulle, Robert J. "Institutionalizing Delay: Foundation Funding and the Creation of US Climate Change Counter-Movement Organizations." *Climatic Change* 122, no. 4 (February 2014): 681–94.

Brune, Michael. "From Walden to the White House." *Coming Clean* (blog), January 22, 2013. http://sierraclub.typepad.com/michael brune/2013/01/civil-disobedience.html.

Buell, Lawrence. *The Environmental Imagination: Thoreau, Nature Writing, and the Formation of American Culture*. Cambridge, MA: Harvard University Press, 1995.

Bullard, Robert D. *Dumping in Dixie: Race, Class, and Environmental Quality*. 3rd ed. Boulder, CO: Westview Press, 2000.

———. *Invisible Houston: The Black Experience in Boom and Bust*. College Station: Texas A&M University Press, 1987.

Bullard, Robert D., Glenn S. Johnson, Denae W. King, and Angel O. Torres. *Environmental Justice Milestones and Accomplishments: 1964–2014.* Houston: Barbara Jordan-Mickey Leland School of Public Affairs, Texas Southern University, Houston, TX, February 2014. http://www.tsu.edu/academics/colleges__schools/publicaffairs/files/pdf/EJMILESTONES2014.pdf.

Bullard, Robert D., and Beverly Wright. *The Wrong Complexion for Protection: How the Government Response to Disaster Endangers African American Communities.* New York: New York University Press, 2012.

Carbon Tracker Initiative. *Unburnable Carbon: Are the World's Financial Markets Carrying a Carbon Bubble?* March 2012. http://www.carbontracker.org/wp-content/uploads/2014/09/Unburnable-Carbon-Full-rev2–1.pdf.

Childress, Kyle. "Proper Work: Wendell Berry and the Practice of Ministry." In *Wendell Berry and Religion: Heaven's Earthly Life*, edited by Joel James Shuman and L. Roger Owens, 71–82. Lexington: University Press of Kentucky, 2009.

———. "Protesters in the Pews." *Christian Century*, January 2, 2013.

Davis, David Brion. *Inhuman Bondage: The Rise and Fall of Slavery in the New World.* New York: Oxford University Press, 2006.

———. *The Problem of Slavery in the Age of Emancipation.* New York: Knopf, 2014.

———. "Should You Have Been an Abolitionist?" *New York Review of Books*, June 21, 2012.

DeChristopher, Tim. "Power Shift 2011 Keynote." YouTube video. April 17, 2011. http://www.youtube.com/watch?v=81EZUkYzrxU.

———. Statement at sentencing, Federal Court, Salt Lake City, Utah, July 26, 2011. http://www.peacefuluprising.org/tims-official-statement-at-his-sentencing-hearing-20110726.

———. Speech presented at "Transformation without Apocalypse: How to Live Well on an Altered Planet," Spring Creek Project, Oregon State University, Corvallis, OR, February 14–15, 2014. http://liberalarts.oregonstate.edu/video/transformation-without-apocalypse-episode-6-tim-dechristopher.

Deep Decarbonization Pathways Project. *Pathways to Deep Decarbonization.* Sustainable Development Solutions Network and Institute for Sustainable Development and International Relations: September 2014. http://unsdsn.org/wp-content/uploads/2014/09/DDPP_Digit_updated.pdf

Delbanco, Andrew. *The Abolitionist Imagination.* Foreword by Daniel Carpenter. Cambridge, MA: Harvard University Press, 2012.

DeWitt, Calvin B. *Earth-Wise: A Biblical Response to Environmental Issues.* 2nd ed. Grand Rapids, MI: Faith Alive Christian Resources, 2007.

Douglass, Frederick. *My Bondage and My Freedom.* Edited by John David Smith. New York: Penguin, 2003.

———. *Narrative of the Life of Frederick Douglass, an American Slave, Written by Himself.* Cambridge, MA: Harvard University Press, 1960.

———. *Selected Speeches and Writings.* Edited by Philip S. Foner. Abridged and adapted by Yuval Taylor. Chicago: Lawrence Hill Books, 1999.

Emanuel, Kerry. *What We Know About Climate Change.* 2nd ed. Cambridge, MA: MIT Press, 2012.

Emerson, Ralph Waldo. *The Essential Writings of Ralph Waldo Emerson.* Edited by Brooks Atkinson. New York: Modern Library, 2000.

Epstein, Paul, and Dan Ferber. *Changing Planet, Changing Health: How the Climate Crisis Threatens Our Health and What We Can Do About It.* Berkeley: University of California Press, 2011.

Evangelical Climate Initiative. *Climate Change: An Evangelical Call to Action.* February 2006. http://www.npr.org/documents/2006/feb/evangelical/callto action.pdf.

First National People of Color Environmental Leadership Summit. *Principles of Environmental Justice.* Washington, DC: October 24–27, 1991. http://www.ejnet.org/ej/principles.html.

Foner, Eric. *The Fiery Trial: Abraham Lincoln and American Slavery.* New York: W. W. Norton, 2010.

———. *Gateway to Freedom: The Hidden History of the Underground Railroad.* New York: W. W. Norton, 2015.

Gage, Beth, and George Gage. *Bidder 70*. Telluride, CO: Gage & Gage Productions, 2012. http://www.bidder70film.com/.

Gandhi, Mohandas K. *An Autobiography: My Experiments with Truth*. Boston: Beacon Press, 1957.

————. *Hind Swaraj and Other Writings*. Cambridge, UK: Cambridge University Press, 1997.

Ganz, Marshall. *Why David Sometimes Wins: Leadership, Organization, and Strategy in the California Farm Worker Movement*. New York: Oxford University Press, 2009.

Gelbspan, Ross. *Boiling Point: How Politicians, Big Oil and Coal, Journalists, and Activists Are Fueling the Climate Crisis—and What We Can Do to Avert Disaster*. New York: Basic Books, 2005.

————. *The Heat Is On: The Climate Crisis, the Cover-Up, the Prescription*. New York: Basic Books, 1998.

Genoways, Ted. "Port Arthur, Texas: American Sacrifice Zone." *On Earth*, August 26, 2013.

Gilding, Paul. *The Great Disruption: Why the Climate Crisis Will Bring on the End of Shopping and the Birth of a New World*. New York: Bloomsbury, 2011.

Gillis, Justin. "A Tricky Transition from Fossil Fuel: Denmark Aims for 100 Percent Renewable Energy." *New York Times*, November 10, 2014.

Goodell, Jeff. "Meet America's Most Creative Climate Criminal." *Rolling Stone*, July 7, 2011.

Gottlieb, Roger S. *A Greener Faith: Religious Environmentalism and Our Planet's Future*. New York: Oxford University Press, 2006.

Gregg, Richard. *The Power of Nonviolence*. Canton, ME: Greenleaf Books, 1960.

Gura, Philip F. *American Transcendentalism: A History*. New York: Hill & Wang, 2007.

Halberstam, David. *The Children*. New York: Random House, 1998.

Hamm, Thomas D. *The Quakers in America*. New York: Columbia University Press, 2003.

Hansen, James. *Storms of My Grandchildren: The Truth About the Coming Climate Catastrophe and Our Last Chance to Save Humanity*. New York: Bloomsbury, 2009.

Hansen, James, et al. "Assessing 'Dangerous Climate Change': Required Reduction of Carbon Emissions to Protect Young People, Future Generations and Nature." *PLOS ONE* 8, no. 12 (December 3, 2013): e81648. doi:10.1371/journal.pone.0081648.

Hansen, James, and Pushker Kharecha. "Assessing 'Dangerous Climate Change': Required Reduction of Carbon Emissions to Protect Young People, Future Generations and Nature." Summary sent to James Hansen's e-mail list, December 3, 2013. http://www.columbia.edu/~jeh1/mailings/2013/20131202_PopularSciencePlosOneE.pdf.

Hare, Bill. *Fossil Fuels and Climate Protection: The Carbon Logic.* Greenpeace International, 1997. http://www.greenpeace.org/international/Global/international/planet-2/report/2006/3/fossil-fuels-and-climate-prote.pdf.

Hawken, Paul. *Blessed Unrest: How the Largest Movement in the World Came into Being and Why No One Saw It Coming.* New York: Viking Penguin, 2007.

Hayhoe, Katharine, and Andrew Farley. *A Climate for Change: Global Warming Facts for Faith-Based Decisions.* New York: FaithWords, 2009.

Hertsgaard, Mark. *Earth Odyssey: Around the World in Search of Our Environmental Future.* New York: Broadway, 1998.

———. "The End of the Arctic?" *Daily Beast*, December 13, 2013.

———. *Hot: Living Through the Next Fifty Years on Earth.* Boston: Houghton Mifflin Harcourt, 2011.

Hochschild, Adam. *Bury the Chains: Prophets and Rebels in the Fight to Free an Empire's Slaves.* Boston: Houghton Mifflin, 2005.

Hodder, Alan D. *Thoreau's Ecstatic Witness.* New Haven, CT: Yale University Press, 2001.

Hopkins, Rob. *The Transition Handbook: From Oil Dependency to Local Resilience.* Cambridge, UK: Green Books, 2008.

Intergovernmental Panel on Climate Change (IPCC). *Climate Change 2007: Synthesis Report.* Geneva: IPCC, 2007.

———. *Climate Change 2014: Synthesis Report.* Geneva: IPCC, 2014.

International Energy Agency. *World Energy Outlook 2011.* November 2, 2011. http://www.worldenergyoutlook.org/publications/weo-2011/.

————. *World Energy Outlook 2012*. November 12, 2012. http://www
.worldenergyoutlook.org/publications/weo-2012/.

————. *World Energy Outlook 2013*. November 12, 2013. http://www
.worldenergyoutlook.org/publications/weo-2013/.

————. *World Energy Outlook 2014*. November 12, 2014. http://www
.world energyoutlook.org/publications/weo-2014/.

Jacobson, Mark Z., and Mark A. Delucchi. "A Plan to Power 100 Percent of the Planet with Renewables." *Scientific American*, November
2009.

————. "Providing All Global Energy with Wind, Water, and Solar
Power, Part I: Technologies, Energy Resources, Quantities and Areas
of Infrastructure, and Materials." *Energy Policy* 39 (2011): 1154–69.
http://web.stanford.edu/group/efmh/jacobson/Articles/I/JDEn
PolicyPt1.pdf.

Jamieson, Dale. *Reason in a Dark Time: Why the Struggle Against Climate Change Failed—and What It Means for Our Future*. New York:
Oxford University Press, 2014.

Joseph, Peniel E. *Stokely: A Life*. New York: Basic Civitas, 2014.

————. *Waiting 'til the Midnight Hour: A Narrative History of Black
Power in America*. New York: Henry Holt, 2006.

Kazin, Michael. *American Dreamers: How the Left Changed a Nation*.
New York: Knopf, 2011.

Kelley, Hilton. *A Lethal Dose of Smoke and Mirrors: Going Home for
Better or Worse*. Port Arthur, TX: Convibe, 2014.

King, Martin Luther, Jr. *A Call to Conscience: The Landmark Speeches
of Dr. Martin Luther King, Jr.* Edited by Clayborne Carson and Kris
Shepard. New York: Hachette, 2001.

————. *A Gift of Love: Sermons from "Strength to Love" and Other
Preachings*. Boston: Beacon Press, 2012.

————. *The Radical King*. Edited with an introduction by Cornel
West. Boston: Beacon Press, 2015.

————. *A Testament of Hope: The Essential Writings and Speeches of
Martin Luther King, Jr.* Edited by James M. Washington. New York:
HarperCollins, 1986.

————. *Where Do We Go From Here: Chaos or Community?* Boston: Beacon Press, 2010.

Klein, Naomi. *The Shock Doctrine: The Rise of Disaster Capitalism.* New York: Henry Holt, 2007.

————. *This Changes Everything: Capitalism vs. the Climate.* New York: Simon & Schuster, 2014.

Kolbert, Elizabeth. *Field Notes from a Catastrophe: Man, Nature, and Climate Change.* New York: Bloomsbury, 2006.

————. *The Sixth Extinction: An Unnatural History.* New York: Henry Holt, 2014.

Lelyveld, Joseph. *Great Soul: Mahatma Gandhi and His Struggle with India.* New York: Knopf, 2011.

Lewis, Paul, and Adam Federman. "Revealed: FBI Violated Its Own Rules While Spying on Keystone XL Opponents." *Guardian*, May 12, 2015.

Mann, Michael E. *The Hockey Stick and the Climate Wars.* New York: Columbia University Press, 2012.

McKanan, Dan. *Prophetic Encounters: Religion and the American Radical Tradition.* Boston: Beacon Press, 2011.

McKibben, Bill. *The Comforting Whirlwind: God, Job, and the Scale of Creation.* Cambridge, MA: Cowley, 2005.

————. *Deep Economy: The Wealth of Communities and the Durable Future.* New York: St. Martin's, 2007.

————. *Eaarth: Making a Life on a Tough New Planet.* New York: Times Books, 2010.

————. *The End of Nature.* New York: Random House, 2006.

————. "The Fossil Fuel Resistance." *Rolling Stone*, April 25, 2013.

————. "Global Warming's Terrifying New Math." *Rolling Stone*, August 2, 2012.

————. "God's Taunt." Sermon delivered at the Riverside Church, New York, NY, April 28, 2013. http://www.theriversidechurchny.org /pdfs/Bill_McKibben-Gods_Taunt-28April2013.pdf.

————. *Hope, Human and Wild: True Stories of Living Lightly on the Earth.* Minneapolis: Milkweed Editions, 2007.

———. "Obama and Climate Change: The Real Story." *Rolling Stone*, December 19, 2013–January 2, 2014.

———. *Oil and Honey: The Education of an Unlikely Activist.* New York: Times Books, 2014.

Mitchell, John Hanson. *Ceremonial Time: Fifteen Thousand Years on One Square Mile.* New York: Doubleday, 1984.

———. *Living at the End of Time.* Boston: Houghton Mifflin, 1990.

———. *Walking Towards Walden: A Pilgrimage in Search of Place.* New York: Perseus, 1995.

Mock, Brentin. *Grist* (blog). http://grist.org/author/brentin-mock/.

Mooney, Chris. "The Melting of Antarctica Was Already Really Bad. It Just Got Worse." *Washington Post*, March 16, 2015.

Mooney, Chris, and Joby Warrick. "Research Casts Alarming Light on Decline of West Antarctic Glaciers." *Washington Post*, December 4, 2014.

Moore, Hilary, and Joshua Kahn Russell. *Organizing Cools the Planet: Tools and Reflections to Navigate the Climate Crisis.* Oakland, CA: PM Press, 2011.

NOAA National Climatic Data Center. *State of the Climate: Global Snow & Ice for September 2012.* October 2012. http://www.ncdc.noaa.gov/sotc/global-snow/2012/9.

National Research Council of the National Academies of Science. *America's Climate Choices.* Washington, DC: National Academies Press, 2011.

Nordhaus, Ted, and Michael Shellenberger. *Break Through: From the Death of Environmentalism to the Politics of Possibility.* Boston: Houghton Mifflin, 2007.

———, eds. *Love Your Monsters: Postenvironmentalism and the Anthropocene.* Oakland, CA: Breakthrough Institute, 2011.

Nordhaus, William. *The Climate Casino: Risk, Uncertainty, and Economics for a Warming World.* New Haven, CT: Yale University Press, 2013.

Oates, Stephen B. *Let the Trumpet Sound: A Life of Martin Luther King, Jr.* New York: Harper & Row, 1982.

Oreskes, Naomi, and Erik M. Conway. *The Collapse of Western Civilization: A View from the Future*. New York: Columbia University Press, 2014.

———. *Merchants of Doubt: How a Handful of Scientists Obscured the Truth on Issues from Tobacco Smoke to Global Warming*. New York: Bloomsbury, 2010.

Patterson, Jacqui, Katie Fink, Camille Grant, and Sabrina Terry. *Just Energy Policies: Reducing Pollution and Creating Jobs*. National Association for the Advancement of Colored People (NAACP) Environmental and Climate Justice Program: February 2014. http://naacp.3cdn.net/8654c676dbfc968f8f_dk7m6j5vo.pdf.

Peaceful Uprising. "Tim's Story." http://www.peacefuluprising.org/tim-dechristopher/tims-story.

Petrulionis, Sandra Harbert. *To Set This World Right: The Antislavery Movement in Thoreau's Concord*. Ithaca, NY: Cornell University Press, 2006.

Pooley, Eric. *The Climate War: True Believers, Power Brokers, and the Fight to Save the Earth*. New York: Hyperion, 2010.

Posner, Eric, and David Weisbach. *Climate Change Justice*. Princeton, NJ: Princeton University Press, 2010.

PricewaterhouseCoopers (PwC). *Too Late for Two Degrees? Low Carbon Economy Index 2012*. PwC: November 2012. http://www.pwc.com/en_GX/gx/sustainability/publications/low-carbon-economy-index/assets/pwc-low-carbon-economy-index-2012.pdf.

Primack, Richard B. *Walden Warming: Climate Change Comes to Thoreau's Woods*. Chicago: University of Chicago Press, 2014.

Ransby, Barbara. *Ella Baker and the Black Freedom Movement: A Radical Democratic Vision*. Chapel Hill: University of North Carolina Press, 2003.

Revkin, Andrew. *Dot Earth* (blog). http://dotearth.blogs.nytimes.com/.

Reynolds, David S. *John Brown, Abolitionist: The Man Who Killed Slavery, Sparked the Civil War, and Seeded Civil Rights*. New York: Knopf, 2005.

Richardson, Robert D. *Henry Thoreau: A Life of the Mind*. Berkeley: University of California Press, 1986.

Roberts, David. *Grist* (blog). http://grist.org/author/david-roberts/.

———. "The Brutal Logic of Climate Change." *Grist* (blog), December 6, 2011. http://grist.org/climate-change/2011–12–05-the-brutal-logic-of-climate-change/.

———. "Yes, We Can Beat Climate Change—But It Will Take Massive International Government Coordination." *Grist* (blog), November 24, 2014. http://grist.org/climate-energy/yes-we-can-beat-climate-change-but-it-will-take-massive-international-government-coordination/.

Romm, Joe. *Climate Progress* (blog). http://thinkprogress.org/person/joe/.

———. "Global Warming Boosts Chances of Civilization-Threatening Megadroughts Here and Abroad." *Climate Progress* (blog), September 4, 2014. http://thinkprogress.org/climate/2014/09/04/3478274/global-warming-megadroughts/.

———. "The Next Dust Bowl." *Nature*, October 27, 2011.

Schewe, Jacob et al. "Multimodel Assessment of Water Scarcity Under Climate Change." Edited by Hans Joachim Schellnhuber. *PNAS* 111, no. 9 (March 4, 2014): 24344289. doi:10.1073/pnas.1222460110.

Shue, Henry. *Climate Justice: Vulnerability and Protection.* New York: Oxford University Press, 2014.

Skocpol, Theda. "Naming the Problem: What It Will Take to Counter Extremism and Engage Americans in the Fight against Global Warming." Paper presented at the Politics of America's Fight against Global Warming symposium, Harvard University, Cambridge, MA, February 14, 2013.

Snyder, Gary. *Danger on Peaks.* Washington, DC: Shoemaker & Hoard, 2004.

———. *Mountains and Rivers Without End.* Berkeley, CA: Counterpoint, 1996.

———. *No Nature: New and Selected Poems.* New York: Pantheon, 1993.

———. *The Practice of the Wild: Essays.* San Francisco: North Point, 1990.

————. *Riprap and Cold Mountain Poems: Fiftieth Anniversary Edition.* Berkeley, CA: Counterpoint, 2009.

Solnit, Rebecca. *A Paradise Built in Hell: The Extraordinary Communities That Arise in Disaster.* New York: Viking Penguin, 2009.

Speth, James Gustave. *America the Possible: Manifesto for a New Economy.* New Haven, CT: Yale University Press, 2012.

————. *The Bridge at the Edge of the World: Capitalism, the Environment, and Crossing from Crisis to Sustainability.* New Haven, CT: Yale University Press, 2008.

Stephenson, Wen. "A Convenient Excuse: Dear Mainstream Media Colleagues: Time Is Running Out to Prevent Climate Catastrophe. Lives Are at Stake. And You Are Failing Us All." *Boston Phoenix,* November 2, 2012.

Stockman, Lorne. "IEA Acknowledges Fossil Fuel Reserves Climate Crunch." *Oil Change International,* November 12, 2012.

Thoreau, Henry David. *Collected Essays and Poems.* New York: Library of America, 2001.

————. *The Essays of Henry D. Thoreau.* Edited by Lewis Hyde. New York: North Point Press, 2002.

————. *Letters to a Spiritual Seeker.* Edited by Bradley P. Dean. New York: W. W. Norton, 2004.

————. *The Maine Woods.* Princeton, NJ: Princeton University Press, 2004.

————. *Walden.* With an introduction and annotations by Bill McKibben. Boston: Beacon Press, 2004.

————. *Walden: A Fully Annotated Edition.* Edited by Jeffrey S. Cramer. New Haven, CT: Yale University Press, 2004.

————. *Walden and Other Writings.* New York: Modern Library, 1981.

Turner, Jack, ed. *A Political Companion to Henry David Thoreau.* Lexington: University Press of Kentucky, 2009.

US Global Change Research Program. *Climate Change Impacts in the United States: The Third National Climate Assessment.* Washington, DC: 2014. http://s3.amazonaws.com/nca2014/high/NCA3_Climate_Change_Impacts_in_the_United%20States_HighRes.pdf.

Ward, Ken. "Bright Lines." *Grist*, February 7–May 8, 2007. http:// grist.org/series/2007–02–07-bright-lines/.

Ward, Ken, and Jay O'Hara. "Coal Is Stupid: Why We Seek to Close Brayton Point." Lobster Boat Blockade. May 15, 2013. http://lobster boatblockade.org/the-action/coal-is-stupid-manifesto/.

Wilkinson, Katharine K. *Between God and Green: How Evangelicals Are Cultivating a Middle Ground on Climate Change.* New York: Oxford University Press, 2012.

Williams, Terry Tempest. "What Love Looks Like: A Conversation with Tim DeChristopher." *Orion*, January/February 2012.

Wilson, E. O. *The Creation: An Appeal to Save Life on Earth.* New York: W. W. Norton, 2006.

———. "Prologue: A Letter to Thoreau." In *The Future of Life*, xi–xxiv. New York: Knopf, 2002.

The World Bank. *Turn Down the Heat: Why a 4C Warmer World Must Be Avoided.* A report for the World Bank by the Potsdam Institute for Climate Impact Research and Climate Analytics. Washington, DC: The World Bank, 2012.

Zelko, Frank. *Make It a Green Peace! The Rise of Countercultural Environmentalism.* New York: Oxford University Press, 2013.

Zinn, Howard. *SNCC: The New Abolitionists.* Boston: Beacon Press, 1964.